The Art of Mathe...

T0093518

Lovers of mathematics, young and old, p ... enjoy this book. It is mathematics with fun: a collection of attractive problems that will delight and test readers. Many of the problems are drawn from the large number that have entertained and challenged students, guests and colleagues over the years during afternoon tea. The problems have their roots in many areas of mathematics. They vary greatly in difficulty: some are very easy, but most are far from trivial, and quite a few rather hard. Many provide substantial and surprising results that form the tip of an iceberg, providing an introduction to an important topic.

To enjoy and appreciate the problems, readers should browse the book, choose one that looks particularly enticing, and think about it on and off for a while before resorting to the hint or the solution.

Follow threads for an enjoyable and enriching journey through mathematics.

BÉLA BOLLOBÁS has been a Fellow at Trinity College, Cambridge, for over fifty years, for decades as a Director of Studies in Mathematics, teaching the very best undergraduates in England, and is the Chair of Excellence in Combinatorics at the University of Memphis. He has had over seventy Ph.D. students. He is a Fellow of the Royal Society and a Member of the Academia Europaea, and a Foreign Member of the Hungarian Academy of Sciences and of the Polish Academy of Sciences. Among the awards he has received are a Senior Whitehead Prize (2007), a Bocskai Prize (2016), a Széchenyi Prize (2017) and an Honorary Doctorate from Adam Mickiewicz University, Poznań. This is his thirteenth book.

Learning is the greatest pleasure in life

The Art of Mathematics – Take Two
Tea Time in Cambridge

Béla Bollobás
University of Cambridge and
University of Memphis

CAMBRIDGE
UNIVERSITY PRESS

University Printing House, Cambridge CB2 8BS, United Kingdom

One Liberty Plaza, 20th Floor, New York, NY 10006, USA

477 Williamstown Road, Port Melbourne, VIC 3207, Australia

314–321, 3rd Floor, Plot 3, Splendor Forum, Jasola District Centre, New Delhi – 110025, India

103 Penang Road, #05–06/07, Visioncrest Commercial, Singapore 238467

Cambridge University Press is part of the University of Cambridge.

It furthers the University's mission by disseminating knowledge in the pursuit of education, learning, and research at the highest international levels of excellence.

www.cambridge.org
Information on this title: www.cambridge.org/9781108833271
DOI: 10.1017/9781108973885

First published 2022

Printed in the United Kingdom by TJ Books Limited, Padstow Cornwall

A catalogue record for this publication is available from the British Library.

ISBN 978-1-108-83327-1 Hardback
ISBN 978-1-108-97826-2 Paperback

Contents

v

Preface

This collection of problems is a sequel to *The Art of Mathematics – Coffee Time in Memphis* (CTM). It is a playful tribute to four giants of mathematics and physics I was fortunate to know well: Paul Erdős, Paul Adrien Maurice Dirac, Israil Moiseevich Gelfand and John Edensor Littlewood. As in CTM, many of the problems in this volume are the kind they would have liked to think about. There are also echoes of the influence of my very early mathematical gurus, Baron Gábor Splényi and the geometer István Reiman.

I was horrified when I realized that as I have grown older, I have completely forgotten a number of the gems of classical elementary mathematics I had known so well in my early teens. It is for this reason that several of those gems have found their way into this collection. These ought to be well known by most people interested in mathematics.

This is not a volume for systematic study, but a book to enjoy. The problems have been selected for their beauty and the elegance of their solutions. Questions on the same topic are not collected into separate chapters, as I wanted to avoid the impression that this book can be used as a sound introduction to various topics. Rather, my hope is that the questions will whet the appetite of readers by giving them food for thought without much previous work.

Who are my intended readers? I have tried to make this volume appeal to people with vastly different backgrounds: students who enjoy doing mathematics, professional mathematicians looking for some relaxation, and also all who loved mathematics in their younger days and are still happy to think about it. Even more, I hope that just about everyone in academia will benefit from dipping into this volume.

Some of the problems are very easy, but others are likely to be pretty demanding even for excellent mathematicians. My hope is that a reader will be fascinated by a problem or two, and will be happy to turn them over in his head whether progress is forthcoming or not. I have always found that it is most pleasant to have a problem on my mind that I cannot do. Most problems have 'hints' that should give some help to the reader without taking away the pleasure of finding a complete solution.

The selection of problems shows a bias towards Cambridge mathematicians and, within Cambridge, towards members of Trinity College. As I have been a Fellow of Trinity College for over fifty years, I hope that this bias can be forgiven. My great respect for Trinity College was instilled in me close to sixty years ago by the Trinity mathematicians Harold Davenport and J.E. Littlewood, and the physicist Paul Dirac, who was actually a Fellow of St John's College.

My chief desire has been to make the book readable; in particular, in the proofs I have not aimed for brevity: I often remind the reader of the relevant definitions and facts, so that he does not have to rack his brain to continue the proof. For this reason mathematicians are likely to find that the proofs go too slowly, but less experienced readers might welcome proofs spelled out in full!

The structure of this book is identical with that of CTM: the first section has the Problems, the second the Hints, and the third the Solutions, i.e. the proofs of the appropriate assertions. Needless to say, the reader should try to solve a problem without reading its hint, and get that help only when he is in dire need of it.

Most of the solutions are followed by notes which tend to be longer than in CTM, as they contain remarks not only about the mathematics, but also about the mathematicians involved with the problems.

I would be disappointed to learn that some people may read one of the problems, ponder about it for a minute or two, and then go on to read its solution. That would be a complete misuse of this volume, like hammering in a nail with a screwdriver. If a problem is found to need more mathematical expertise than the reader has, I would recommend abandoning that problem until such time as he has acquired the relevant background: this volume has plenty of problems that need little mathematical sophistication.

There are many people who drew my attention to beautiful problems, and gave me the pleasure of discussing those problems with them. I have received especially much help from Paul Balister (Oxford) and Imre Leader (Cambridge): I owe them a great debt of gratitude. I am also grateful to Józsi Balogh (Urbana), Enrico Bombieri (IAS, Princeton), Tim Gowers (Cambridge), Andrew Granville (Montreal), Misi Hujter (Budapest), Rob Morris (IMPA, Rio de Janeiro), Julian Sahasrabudhe (Cambridge), Tadashi Tokieda (Stanford) and Mark Walters (London).

This book would not have been finished without the great help I have received from my brilliant friend of many decades, David Tranah, the Editorial Director for Mathematics at Cambridge University Press – much more than I could reasonably have hoped for. From the other side of the pond, my excellent Editorial Assistant, Tricia Simmons, has also given me much help. I am deeply grateful to both of them.

I should like to thank my current research students, Vojtěch Dvořák, Peter van Hintum, Harry Metrebian, Adva Mond, Jan Petr, Julien Portier, Victor Souza and Marius Tiba, for reading parts of the manuscript, and saving me from numerous howlers, like proving the necessity of a condition twice and forgetting to prove its sufficiency. I am sure that many mistakes have remained, for which I apologize.

Finally, this volume would never have been completed without the help and understanding of my wife, Gabriella, who has also put artistic 'Art' into it.

Béla Bollobás
Cambridge, Ascension Day, 2021.

The Problems

1. (Real Sequences) (i) At most how many real numbers can be chosen from the open interval $(0, 2n + 1)$ if none is at distance less than 1 from an integer multiple of another? To spell it out, let $n \geq 1$ be a fixed natural number. Suppose that $0 < x_1 < \cdots < x_N < 2n + 1$ are such that $|kx_i - x_j| \geq 1$ for all natural numbers i, j and k with $1 \leq i < j \leq N$. At most how large is N?

(ii) At most how many real numbers can be chosen from the open interval $(0, (3n + 1)/2) = (0, 3n/2 + 1/2)$ if none is at distance less than 1 from an *odd* multiple of another?

2. (Vulgar Fractions) Show that every rational number r, $0 < r < 1$, is the sum of a finite number of reciprocals of distinct natural numbers. For example,

$$\frac{4699}{7320} = \frac{1}{2} + \frac{1}{8} + \frac{1}{60} + \frac{1}{3660}.$$

3. (Rational and Irrational Sums) Let $2 \leq n_1 < n_2 < \cdots$ be a sequence of positive integers such that

$$n_{i+1} \geq n_i(n_i - 1) + 1$$

for every $i \geq 1$, and set

$$r = \sum_{i=1}^{\infty} \frac{1}{n_i}.$$

Show that r is rational if and only if $n_{i+1} = n_i(n_i - 1) + 1$ for all but finitely many values of i.

4. (Ships in Fog) Five ships, A, B, C, D and E are sailing in a fog with constant and different speeds, and constant and different straight-line courses, with different directions. The seven pairs AB, AC, AD, BC, BD, CE and DE

have each had near misses, call them '*collisions*'. Does it follow that, in addition, E collides with either A or B? Maybe both? And does C collide with D?

5. (A Family of Intersections)   For $0 < p < 1$, a *p-random subset* $X = X_p$ of $[n] = \{1, 2, \ldots, n\}$ is obtained by taking n independent binomial random variables $\xi_1, \xi_2, \ldots, \xi_n$ with $\mathbb{P}(\xi_i = 1) = p = 1 - \mathbb{P}(\xi_i = 0)$, and setting $X_p = \{i : \xi_i = 1\}$. The probability measure $\mathbb{P}_p$ on $\mathcal{P}_n$, the set of all 2^n subsets of $[n]$, is given by

$$\mathbb{P}_p(A) = \mathbb{P}(X_p = A) = p^{|A|} (1 - p)^{n - |A|},$$

so that the probability of a family $\mathcal{A} \subset \mathcal{P}_n$ is

$$\mathbb{P}_p(\mathcal{A}) = \sum_{A \in \mathcal{A}} \mathbb{P}_p(A) = \sum_{A \in \mathcal{A}} p^{|A|} (1 - p)^{n - |A|}.$$

Let $\mathcal{A} \subset \mathcal{P}_n$ have p-probability r: $\mathbb{P}_p(\mathcal{A}) = r$, and define

$$\mathcal{J} = \mathcal{J}(\mathcal{A}) = \{A \cap B : A, B \in \mathcal{A}\}.$$

Show that

$$\mathbb{P}_{p^2}(\mathcal{J}) \geq r^2.$$

6. (The Basel Problem)   Forget for a moment the mathematical rigour we have to have in our proofs, and give a beautiful solution of the famous 'Basel Problem': prove that

$$\sum_{k=1}^{\infty} 1/k^2 = 1 + \frac{1}{4} + \frac{1}{9} + \frac{1}{16} \cdots = \pi^2/6.$$

7. (Reciprocals of Primes)   Give three proofs of the theorem that the sum of reciprocals of the primes is divergent: $\sum_p 1/p = \infty$, where the summation is over the primes.

8. (Reciprocals of Integers)   Let $1 < n_1 < n_2 < \cdots$ be a sequence of natural numbers such that $\sum_{i=1}^{\infty} 1/n_i < \infty$. Show that the set

$$M = M(n_1, n_2, \ldots) = \{n_1^{\alpha_1} \ldots n_k^{\alpha_k} : \alpha_i \geq 0\}$$

has zero density, i.e. if $\varepsilon > 0$ and n is large enough (depending on ε) then there are at most εn elements of M that are at most n.

9. (Completing Matrices)   For $1 \leq k < n$, let $\mathcal{A}_{k,n}$ be the collection of $n \times n$ matrices with each entry zero or one, having precisely k ones in each row and each column. Show that for $1 \leq r < n$ an $r \times n$ matrix of zeros and ones has an extension to a matrix in $\mathcal{A}_{k,n}$ if and only if each row has precisely k ones, and in each column there are at least $k + r - n$ and at most k ones.

10. (Convex Polyhedra – Take One)   Is there a convex polyhedron which contains a point whose perpendicular projection on the plane of every face falls outside the face? And just fails to fall in the interior of the face?

11. (Convex Polyhedra – Take Two)   Show that every 3-dimensional polyhedron with at least thirteen faces has a face meeting at least six other faces. Two faces are said to meet if they share a vertex or an edge.

12. (A Very Old Tripos Problem)   Let p, q and r be complex numbers with $pq \neq r$. Transform the cubic $x^3 - px^2 + qx - r = 0$, where the roots are a, b, c, into one whose roots are $\frac{1}{a+b}$, $\frac{1}{a+c}$, $\frac{1}{b+c}$.

13. (Angle Bisectors)   Show that if two (internal) angle bisectors of a triangle are equal then the angles themselves are also equal.

14. (Chasing Angles – Take One)   Let ABC be an isosceles triangle with angle $20°$ at the apex A. Let D be a point on AB and E a point on AC such that $\angle BCD = 50°$ and $\angle CBE = 60°$, as in Figure 1. What is the angle $\angle BED$?

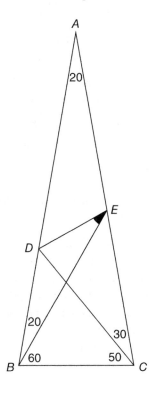

Figure 1 Adventitious angles

15. (Chasing Angles – Take Two)   Let ABC be an isosceles triangle with angle $20°$ at the apex A and so angles $80°$ at the base; furthermore, let D be a point on the side AB such that $\measuredangle CD = 10°$, and E on AC such that $\measuredangle BE = 20°$, as in Figure 2. Use entirely elementary methods, without any recourse to trigonometry, to determine the angle $\measuredangle DE$.

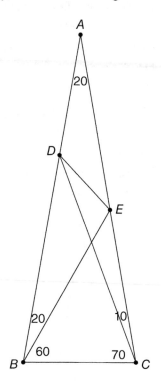

Figure 2 Information about our points

16. (Pythagorean Triples)   We call a triple (a, b, c) of natural numbers a *Pythagorean triple* if $a^2 + b^2 = c^2$. Also, a Pythagorean triple (a, b, c) is *primitive* or *relatively prime* if a, b and c do not have a common divisor (greater than 1). Clearly, every Pythagorean triple is a multiple of a primitive Pythagorean triple. Also, if (a, b, c) is a primitive Pythagorean triple then a and b have opposite parities since if both of them are odd then the sum of their squares is 2 modulo 4, so it cannot be a square, and if both of them are even then the sum of their squares is also even, so c also has to be even. Usually we take a to be odd and b even.

Show that (a, b, c) is a primitive Pythagorean triple with a odd and b even if and only if there are relatively prime numbers $u > v \geq 1$ of opposite parity such that $a = u^2 - v^2$, $b = 2uv$ and $c = u^2 + v^2$. Even more, give two proofs, one algebraic and the other geometric.

17. (Fermat's Theorem for Fourth Powers)   Show that the equation $a^4 + b^4 = c^4$ has no solutions in natural numbers. Putting it slightly differently: if a and b are strictly positive integers then $a^4 + b^4$ cannot be a fourth power.

18. (Congruent Numbers)   A natural number n is said to be *congruent* if there is a right-angled triangle with rational sides, whose area is n. For example, the right-angled triangle with sides $3, 4$ and 5 tells us that 6 is congruent. Show that 1 is not a congruent number.

19. (A Rational Sum)   Find a necessary and sufficient condition for a rational number $s > 1$ that ensures that $\sqrt{s+1} - \sqrt{s-1}$ is also rational, where $\sqrt{\cdot}$ denotes the positive square root.

20. (A Quartic Equation)   Find a large family of integer solutions of

$$A^4 + B^4 = C^4 + D^4. \tag{1}$$

More precisely, look for fairly general polynomials A, B, C and D in $\mathbb{Z}[a, b]$ such that (1) holds. To this end, look for the solution in the form

$$A = ax + c, \qquad B = bx - d,$$
$$C = ax + d, \qquad D = bx + c$$

where a, b, c, d and x are rational numbers. Considering a, b, c and d constant, (1) holds if x satisfies a quartic whose first and last coefficients are 0. Show that with a suitable choice of a, b, c and d the coefficient of x^3 is also 0, and use this to find our polynomials.

21. (Regular Polygons)   Show that, of all polygons of the same number of sides and equal perimeter length, the regular polygon has the greatest area.

22. (Flexible Polygons)   Consider all polygons with given sides but one in a given cyclic order. Show that if the area of such a polygon is maximal then it may have a circle circumscribed about it, having the unknown side for a diameter of the circle.

23. (Polygons of Maximal Area)   Show that the area of a polygon with given sides is not larger than the cyclic polygon with these sides, i.e. the one that may have a circle circumscribed about it. The reader is invited to find several solutions of this problem.

24. (Constructing $\sqrt[3]{2}$)   Let OS_1PS_2 be a $2m \times m$ rectangle with $OS_1 = PS_2$ of length $2m$ and $OS_2 = PS_1$ of length m. Let C be the circle through the vertices O, S_1, P and S_2, and let Q be the point of the PS_2 arc of C such that the line through P and Q meets the (extended) lines OS_1 and OS_2 in R_1 and R_2, and the segments PR_1 and QR_2 have the same length. Finally, let T_1 and T_2 be the projections of Q on the segments OR_1 and OR_2. Show that OT_1 has length $\sqrt[3]{2}m$.

25. (Circumscribed Quadrilaterals)   Let $ABCD$ be a quadrilateral circumscribed about a circle with centre O. Let E and F be the midpoints of the

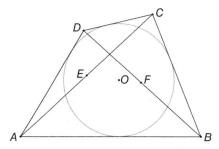

Figure 3 A quadrilateral circumscribed about a circle with centre O; the points E and F are the midpoints of the diagonals AC and BD.

diagonals AC and BD, as in Figure 3. Show that E, F and O are collinear.

26. (Partitions of Integers)   A *partition* of an integer n is a sequence $\lambda = (\lambda_1, \ldots, \lambda_k)$ of positive integers $\lambda_1 \geq \cdots \geq \lambda_k \geq 1$ whose sum is n. This partition is also written as $\lambda_1 + \cdots + \lambda_k$. Each λ_i is a *summand* or *part*; the number of parts, k, is the *length* of the partition λ. As customary, we shall write $p(n)$ for the *partition function*, the number of partitions of $n \geq 1$. Note that 4 has five partitions: $4, 3 + 1, 2 + 2, 2 + 1 + 1, 1 + 1 + 1 + 1$, so $p(4) = 5$, and 5 has seven partitions: $5, 4 + 1, 3 + 2, 3 + 1 + 1, 2 + 2 + 1, 2 + 1 + 1 + 1, 1 + 1 + 1 + 1 + 1$, so $p(5) = 7$. Also, $p(0) = 1$: the only partition of 0 is the empty partition.

(i) Show that the formal power series $\sum_{n=0}^{\infty} p(n)x^n$, called the *generating function* of $p(n)$, is

$$(1 + x + x^2 + x^3 + \cdots)(1 + x^2 + x^4 + x^6 + \cdots)(1 + x^3 + x^6 + x^9 + \cdots) \cdots,$$

i.e.

$$\frac{1}{1-x} \cdot \frac{1}{1-x^2} \cdot \frac{1}{1-x^3} \cdots.$$

(ii) Give three proofs of the assertion that the number of partitions of n

without 1 as a part is $p(n) - p(n - 1)$. For example, 5 has two partitions not containing 1, namely 5 and $3 + 2$, and $p(5) - p(4) = 7 - 5 = 2$.

27. (Parts Divisible by m and $2m$)   Show that the number of partitions of n in which no multiple of m is repeated is equal to the number of partitions of n without a multiple of $2m$.

28. (Unequal vs Odd Partitions)   (i) Show that the number of partitions of n into unequal parts is equal to the number of partitions into odd parts.

(ii) Let $m \geq 1$. Show that the number of partitions of n in which no part is repeated more than m times is equal to the number of partitions in which no part is a multiple of $m + 1$.

29. (Sparse Bases)   A set S of natural numbers has density zero if $S(n)$ tends to zero as n tends to infinity, where $S(n)$ is the number of elements of S not greater than n.

Show that there is a set $S \subset \mathbb{N}$ of density zero such that every positive rational is the sum of a finite number of reciprocals of distinct terms of S.

30. (Sets with Small Pairwise Intersections)   Let $A_1, \ldots, A_m \in [n]^{(r)}$, i.e. let $A_1, \ldots, A_m$ be r-subsets of $[n] = \{1, \ldots, n\}$. Show that if $|A_i \cap A_j| \leq s < r^2/n$ for all $1 \leq i < j \leq m$, then $m \leq n(r - s)/(r^2 - sn)$.

Show also that if r^2/n is an integer and $|A_i \cap A_j| < r^2/n$ for all $1 \leq i < j \leq m$, then $m \leq r - r^2/n + 1 \leq n/4 + 1$.

31. (The Diagonals of Zero–One Matrices)   Given $n \geq 1$, let $\mathcal{A}_n$ be the set of all $n \times n$ matrices with each entry 0 or 1. For $A \in \mathcal{A}_n$, write $\mathcal{A}(A)$ for the set of matrices obtained from A by permuting its rows. Thus if no two rows of A are equal then $\mathcal{A}(A)$ consists of $n!$ matrices. Denote by $d(A)$ the number of different main diagonals of the matrices in $\mathcal{A}(A)$. Determine $d(n) = \max\{d(A) : A \in \mathcal{A}_n\}$.

Note that the total number of main diagonals (with entries 0 and 1) is 2^n, which is much smaller than $n!$, so there is no obvious reason why we could not obtain all 2^n diagonals. For example, if the three rows of our 3×3 matrix are 111, 101 and 001, then the diagonals are 101 (if 111 is kept as the first row), 111, 011 and 001 (for the other orders).

32. (Tromino and Tetronimo Tilings)   An $m \times n$ *board* is an $m \times n$ rectangle made up of mn unit squares called cells; a *deficient $m \times n$ board $m \times n$* board from which a cell has been removed. A *tromino* is the union of three cells sharing a vertex, and a T-*tetromino* is the union of four cells in the shape of a

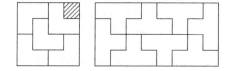

Figure 4 A tromino tiling of a deficient 4×4 board, and a T-tetromino tiling of a 4×8 board.

letter T. We are interested in tilings of a deficient $n \times n$ board by trominoes and an $m \times n$ board by T-tetrominoes, as in Figure 4.

(i) Show that if n is a power of 2 then every deficient $n \times n$ board can be tiled with trominoes.

(ii) Show that if an $m \times n$ rectangle can be tiled with T-tetrominoes then mn is divisible by 8.

33. (Tromino Tilings of Rectangles)   For what values of m and n can an $m \times n$ rectangle be tiled with trominoes, as in Figure 5? [A tromino is a 2×2 square with one quarter cut off, as in Problem 32.]

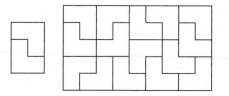

Figure 5 Tromino tilings of a 3×2 board and a 5×9 board.

34. (Number of Matrices)   What is the number of $n \times n$ matrices with non-negative integer entries, in which every row and column has at most three non-zero entries, these non-zero entries are different, and their sum is 7? An example of such a matrix is

$$\begin{pmatrix} 0 & 1 & 0 & 2 & 4 \\ 5 & 0 & 2 & 0 & 0 \\ 0 & 0 & 4 & 0 & 3 \\ 0 & 6 & 0 & 1 & 0 \\ 2 & 0 & 1 & 4 & 0 \end{pmatrix}$$

35. (Halving Circles)   Let S be a set of $2n + 1 \geq 5$ points in the plane in general position. In this context, being in 'general position' means that no three points are on a line and no four points are on a circle. We say that a circle C *halves* S if three points of S are on C, $n-1$ inside C and so $n-1$ outside C. Show that there are at least $n(2n+1)/3$ halving circles.

36. (The Number of Halving Circles) Continuing the previous problem, show that for every $n \geq 1$ there is a set of $2n+1$ points in general position in the plane with exactly n^2 halving circles.

37. (A Basic Identity of Binomial Coefficients) Let $f(X)$ be a polynomial of degree less than n. Show that

$$\sum_{k=0}^{n} (-1)^k \binom{n}{k} f(k) = 0.$$

38. (A Simple Sum?) Put the sum

$$\sum_{i=0}^{n} (-1)^i \binom{n}{i} (x-i)^n$$

into a much simpler form.

39. (Dixon's Identity – Take One) We shall use the convention that $0! = 1$ and $1/k! = 0$ for $k < 0$. Let a, b, c be non-negative integers. Show that

$$\sum_{k} \frac{(-1)^k (a+b)!(b+c)!(c+a)!}{(a+k)!(a-k)!(b+k)!(b-k)!(c+k)!(c-k)!} = \frac{(a+b+c)!}{a!b!c!}.$$

On the left-hand side, the summation is over all integers k; equivalently, we may take the sum $\sum_{-d \leq k \leq d}$, where $d = \min\{a, b, c\}$.

40. (Dixon's Identity – Take Two) (i) Let m and n be non-negative integers, and write X for a variable. Prove the following identity of polynomials with real coefficients:

$$\sum_{k=0}^{2n} (-1)^k \binom{m+2n}{m+k} \binom{X}{k} \binom{X+m}{m+2n-k} = (-1)^n \binom{X}{n} \binom{X+m+n}{m+n}.$$

Here and elsewhere, for a polynomial $f(X)$ over the reals and a non-negative integer ℓ, we write

$$\binom{f(X)}{\ell} = f(X)(f(X)-1)(f(X)-2)\ldots(f(X)-\ell+1).$$

In particular, $\binom{f(X)}{0} = 1$, and if ℓ is a negative integer then $\binom{f(X)}{\ell} = 0$.
(ii) Deduce that if a, b and c are non-negative integers and, say, $b \leq a, c$, then

$$\sum_{k=-b}^{b} (-1)^k \binom{a+b}{a+k} \binom{b+c}{b+k} \binom{c+a}{c+k} = \frac{(a+b+c)!}{a!b!c!}.$$

41. (An Unusual Inequality)   Let $x_0 = 0 < x_1 < x_2 < \cdots$. Show that

$$\sum_{n=1}^{\infty} \frac{x_n - x_{n-1}}{x_n^2 + 1} < \frac{\pi}{2}.$$

42. (Hilbert's Inequality)   Let $(a_n)_1^{\infty}$ and $(b_n)_1^{\infty}$ be square-summable sequences of real numbers: $\sum_n a_n^2 < \infty$ and $\sum_n b_n^2 < \infty$. Show that

$$\sum_{m,n} \frac{a_m b_n}{m+n} < \pi \sqrt{\sum_m a_m^2} \sqrt{\sum_n b_n^2}.$$

43. (The Size of the Central Binomial Coefficient)   Let $k \geq 1$ be an integer and $c, d > 0$ positive real numbers such that

$$\frac{c}{\sqrt{k-1/2}} \, 4^k \leq \binom{2k}{k} \leq \frac{d}{\sqrt{k+1/2}} \, 4^k.$$

Show that then the analogous inequalities hold for all $n \geq k$:

$$\frac{c}{\sqrt{n-1/2}} \, 4^n \leq \binom{2n}{n} \leq \frac{d}{\sqrt{n+1/2}} \, 4^n$$

whenever $n \geq k$. In particular,

$$\binom{2n}{n} < \begin{cases} 2^{2n-1} & \text{if } n \geq 2, \\ 2^{2n-2} & \text{if } n \geq 5, \end{cases}$$

and

$$\frac{0.5}{\sqrt{n-1/2}} \, 4^n \leq \binom{2n}{n} \leq \frac{0.6}{\sqrt{n+1/2}} \, 4^n$$

for $n \geq 4$.

44. (Properties of the Central Binomial Coefficient)   Consider the prime factorization of the central binomial coefficient $\binom{2n}{n}$ for $n \geq 1$:

$$\binom{2n}{n} = \prod_{p<2n} p^{\alpha_p},$$

where p denotes a prime. Show the following assertions:

(i) $\alpha_p = 0$ or 1 if $\sqrt{2n} < p < 2n$;
(ii) $\alpha_p = 0$ if $2n/3 < p \leq n$;
(iii) $p^{\alpha_p} \leq 2n$ for every p.

45. (Products of Primes) For a real number $n \geq 2$, denote by

$$\Pi(n) = \prod_{p \leq n} p$$

the product of all primes that are at most n. (As usual, in our notation p is always a prime.) Show that

$$\Pi(n) < 2^{2n-3}$$

for all $n \geq 2$. Sharpening this inequality a little, show that for $n \geq 9$ we have

$$\Pi(n) < 4^n / n.$$

46. (The Erdős Proof of Bertrand's Postulate) Show that for every $n \geq 1$ there is a prime between n and $2n$: more precisely, there is a prime p satifying $n < p \leq 2n$.

47. (Powers of 2 and 3) Show that no perfect powers of 2 and 3 differ by exactly 1, except 2 and 3, 4 and 3, and 8 and 9.

48. (Powers of 2 Just Less Than a Perfect Power) Show that the only solution of $2^m = r^n - 1$ in positive integers m, r, n greater than 1 is $m = 3, r = 3$ and $n = 2$, giving $2^3 = 3^2 - 1$.

49. (Powers of 2 Just Greater Than a Perfect Power) Show that the equation $2^m = r^n + 1$ has no solutions in positive integers m, r, n greater than 1.

50. (Powers of Primes Just Less Than a Perfect Power) Let $p \geq 3$ be a prime. Show that the equation $p^m = r^n - 1$ has no solution in positive integers m, r, n greater than 1.

51. (Banach's Matchbox Problem) An inveterate smoker puts two boxes of matches into the pockets of his jacket. Every time he wants to light up, he is equally likely to reach into either pocket. After a while, when he takes out one of the boxes, he finds it empty. Having started with n matches in each box, what is the probability that at that moment the other box has k matches?

Not unreasonably, the problem assumes that when the absent-minded smoker uses up the last match of a box, he puts the empty box back into his pocket.

52. (Cayley's Problem) How many convex k-gons can be formed by the diagonals of a convex n-gon?

53. (Min vs Max) Let K_n be a complete graph of order n with non-negative weights on the edges. Define the weight of a subgraph to be the sum of the weights of its edges. Consider two 'natural' ways of constructing a Hamilton path of K_n, i.e. a path through all n vertices. First, starting at a vertex a, always choose the edge of maximal weight continuing the path constructed so far, joining the current end-vertex to a vertex not on the current path. Second, starting from a vertex b, always choose the edge of minimal weight. Show that the weight of the first Hamilton path is at least as large as the weight of the second.

54. (Sums of Squares)   Prove that, if any three numbers are taken that cannot be arranged in an arithmetic progression, and whose sum is a multiple of 3, then the sum of their squares is also the sum of another set of 3 squares, the two sets having no common term.

55. (The Monkey and the Coconuts)   Five men and a monkey were shipwrecked on a desert island. They spent the first day gathering coconuts, which they piled up in a heap before going to sleep. The heap was very big, but could not possibly have contained more than ten thousand coconuts.

In the middle of the night one of the men woke up, and to make sure that he was not shortchanged, divided the coconuts into five equal piles, with a single coconut remaining. He gave that remainder to the monkey, hid his fifth, rearranged the rest into a single heap, and went back to sleep. Later a second man woke up and did the same, then the third, the fourth and the fifth. In the morning the men successfully divided the remaining coconuts into five equal piles: this time no coconuts were left over.

How many coconuts were there in the beginning?

56. (Complex Polynomials)   Given a polynomial h with complex coefficients, write $S_h \subset \mathbb{C}$ for the region where $|h(z)| \leq 1$, so f is 'small'. Show that if f and g, $f \neq g$, are monic polynomials of degree at least one, i.e. $f(z) = z^n + a_1 z^{n-1} + \cdots + a_n$ and $g(z) = z^m + b_1 z^{m-1} + \cdots + b_m$ with $n, m \geq 1$, then S_f cannot be a proper subset of S_g.

57. (Gambler's Ruin)   Rosencrantz and Guildenstern gamble by repeatedly tossing a biased coin with probability p of heads and probability $q = 1 - p$ of tails, where $0 < p < 1$. Every time the coin comes up heads, Rosencrantz wins a krone from Guildenstern, otherwise Guildenstern wins a krone from Rosencrantz: they play till one of them loses all his money, i.e. 'gets ruined' and so the other 'wins'. Starting with the same sum, say, each with k kroner, they play till one of them is ruined. When is the game expected to be longer: if Rosencrantz wins or Guildenstern?

58. (Bertrand's Box Paradox)   There are three identical boxes, with identical drawers on opposite sides, and a coin in each drawer. One of the boxes contains two gold coins, another two silver coins, and the third one of each. We pick a box and a drawer at random, and find a gold coin. What is the probability that the other coin in the box is also gold?

59. (The Monty Hall Problem)   It's 'Let's Make a Deal' – a famous TV show starring Monty Hall.

A contestant is shown three boxes and is told that one of the boxes contains the keys to a new Lincoln Continental, while the other two are empty. If the contestant chooses the box with the keys, he wins the car. Monty asks the contestant to choose a box: the contestant chooses box A, but *does not open it*. Then Monty, who knows perfectly well which box contains the car keys, opens one of the other two boxes, say, box B, and shows to the contestant that it is empty. And now comes the crunch: he asks the contestant whether he would like to swap his box for the third box, box C. Should the contestant swap or stay with his original box A to optimize his chances of winning the Lincoln Continental?

60. (Divisibility in a Sequence of Integers)   Let $a_1 < a_2 < \cdots$ be an infinite sequence of natural numbers. Show that there exists either an infinite subsequence in which no integer divides another, or an infinite subsequence in which every term divides all subsequent terms.

61. (The Moving Sofa Problem)   A long passage of unit width has a right-angled bend in it. A flat rigid plate (made up of one piece) of area A can be manœuvered from one end of the passage to the other. [To spice up the formulation, it is customary to think of the 'flat rigid plate' as a sofa that is being pushed along the floor without lifting or tilting it – hence the traditional colourful name: 'the moving sofa problem'.] Prove that $A < 2\sqrt{2} \sim 2.8284$. Show also that, if the plate has a suitable shape (to be determined), one may have

$$A = \frac{\pi}{2} + \frac{2}{\pi} \sim 2.2074.$$

62. (Minimum Least Common Multiple)   Let $a_1 < a_2 < \cdots < a_n \le 2n$ be a sequence of $n \ge 5$ positive integers. Show that there are integers $a_i < a_j$ such that their least common multiple is at most $6(\lfloor n/2 \rfloor + 1)$:

$$[a_i, a_j] \le 6(\lfloor n/2 \rfloor + 1).$$

Show also that this inequality is best possible.

63. (Vieta Jumping)   Let a and b be positive integers such that $q = (a^2 + b^2)/(ab + 1)$ is also an integer. Show that q is a perfect square.

64. (Infinite Primitive Sequences)   A sequence A of natural numbers is said to be *primitive* if no member of it is a multiple of another. Show that if A is an infinite primitive sequence then

$$\sum_{a \in A} \frac{1}{a} \prod_{p \le p_a} \left(1 - \frac{1}{p}\right) \le 1,$$

where p_a is the largest prime factor of a and p denotes a prime. If $p_a = 2$, i.e. a is a power of 2, then the empty product above is taken to be 1, as always.

65. (Primitive Sequences with a Small Term)   We know from Problem 1 that if $(a_i)_1^\ell$ is a primitive sequence of natural numbers such that $1 < a_1 < a_2 < \cdots < a_\ell \le 2n$ then $\ell \le n$. This upper bound n is trivially best possible, as shown by the primitive sequence $a_1 = n + 1 < a_2 = n + 2 < \cdots < a_n = 2n$. This example consisting only of large numbers begs the question: can a_1 be made much smaller? Show that $a_1 \ge 2^k$, where k is defined by the inequalities $3^k < 2n < 3^{k+1}$.

66. (Hypertrees)   An *r-hypertree* or simply an *r-tree* is an *r*-uniform hypergraph without isolated vertices such that for some order $E_1, \ldots, E_m$ of its edges,

$$\left| E_{k+1} \cap \left(\bigcup_{i=1}^k E_i \right) \right| = 1$$

for every k, $1 \le k < m$. Thus, every 'edge' consists of r vertices, and in some order of the edges, every edge from the second on has precisely one vertex that belongs to any of the previous edges.

Let G be an r-uniform hypergraph containing no copy of a certain r-tree T with m edges. Show that G is k-colourable for $k = 2(r - 1)(m - 1) + 1$.

67. (Subtrees)   At least how many subtrees (of order at least 1) are in a tree on $n \geq 1$ vertices, and at most how many? Are the extremal trees unique?

68. (All in a Row)   All twenty students in a class are asked to line up in a row, one behind another, so that the first student, standing behind all the others, sees all in the class, the second sees all but the first, the third sees all but the first and second, etc., and the twentieth does not see anyone. The teacher places black and white hats at random on each of the students: say, tosses a fair coin for each student, and if the coin comes down heads, a white hat is placed on the student, and a black hat after tails. The students are challenged to call out one by one (for all to hear) the colour of their hats, starting with the first (who sees all) and ending with the twentieth. After the hats have been placed on the students, there is no conferring, but before the game starts, the students are allowed to agree on a strategy to give them as much chance as possible.

If every student guesses at random (after all, the hats are distributed at random, so what else is there to do?), the probability that they all guess correctly is $(1/2)^{20}$, which is less than $1/1,000,000$. Following an optimal strategy, what is the probability that they all guess correctly?

69. (An American Story)   The governor of a prison announces that his twenty prisoners on Death Row will escape execution if only they pass a simple test. Their names are placed in twenty identical boxes, one in each; then these boxes are closed and arranged at random in a row in a closed room. For the test, the prisoners are to be led into the room one by one, and every one of them is allowed to open no more than twelve of the twenty boxes in search of his name. Every prisoner will have to leave the room as he found it and, having left it, is not allowed any communication with the others.

If every one of the twenty inmates finds his name then they will all be reprieved; however, if even one of them fails then all twenty will be executed. Is there a strategy that gives the prisoners a better than even chance to escape execution?

70. (Six Equal Parts)   Let S be a set of $6k$ points in general position in the plane. (Thus every line contains at most two points of S.) Show that there are three concurrent lines that partition S into six equal parts. To spell it out: there are three lines through the same point such that each of the six open plane segments determined by these lines contains exactly k of the points of S.

71. (Products of Real Polynomials)   Show that the degree of the product of two real polynomials (i.e. polynomials with real coefficients) in several variables is the sum of the degrees, and the degree of the sum of squares of real polynomials is twice the maximal degree of the summands.

72. (Sums of Squares)   (i) Let f be a polynomial in one variable with real coefficients: $f \in \mathbb{R}[X]$. Show that if $f(x) \geq 0$ for every real number x then f is the sum of the squares of two real polynomials: $f = g^2 + h^2$, where $g, h \in \mathbb{R}[X]$.

(ii) Let f be a polynomial in two variables ($f \in \mathbb{R}[X, Y]$) such that $f(x, y) \geq 0$ for all $x, y \in \mathbb{R}$. Does it follow that f is a sum of squares, i.e. there are polynomials $f_1, \ldots, f_k \in \mathbb{R}[X, Y]$ such that $f = f_1^2 + \cdots + f_k^2$?

73. (Diagrams of Partitions)   Show that the number of partitions of n into p parts, with largest part q, is equal to the number of partitions of n into q parts, with largest part p.

74. (Even and Odd Partitions)   (i) Show that the number of partitions of n into an even number of unequal summands is equal to the number of partitions of n into an odd number of unequal summands unless $n = k(3k \pm 1)/2$, in which case the difference is $(-1)^k$. For example, for $n = 11$ there are six 'even partitions': $10 + 1$, 92, 83, 74, 65 and 5321, and six 'odd partitions': 11, 821, 731, 641, 632 and 542, and for $n = 3(3 \cdot 3 - 1)/2 = 12$ (i.e. $k = 3$) there are seven even partitions: $11 + 1$, $10 + 2$, 93, 84, 75, 6321 and 5421, and eight odd partitions: 12, 921, 831, 741, 732, 651, 642 and 543. Thus for $k = 3$ we do have $7 - 8 = -1 = (-1)^{-3}$, as we claim.

(ii) Deduce from this that the infinite product $\prod_{k=1}^{\infty}(1 - x^k)$ has the following expansion:

$$\prod_{k=1}^{\infty}(1 - x^k) = \sum_{-\infty}^{\infty}(-1)^k x^{k(3k-1)/2} = 1 + \sum_{k=1}^{\infty}(-1)^k (x^{k(3k-1)/2} + x^{k(3k+1)/2}).$$

75. (Partitions – Maximum and Parity)   Let $p_{o,e}(n)$ be the number of partitions of n into unequal parts in which the largest part is odd and the number of parts is even. [To get one of these partitions, start with an odd number and finish after an even number of steps.] Define $p_{e,o}(n)$, $p_{e,e}(n)$ and $p_{o,o}(n)$ analogously. Thus $p_{o,e}(12) = 4$ counts $11 + 1, 9 + 3, 7 + 5$ and $5 + 4 + 2 + 1$, $p_{e,o}(12) = 4$ counts $12, 8 + 3 + 1, 6 + 5 + 1$ and $6 + 4 + 2$, $p_{o,o}(12) = 4$ counts $9 + 2 + 1, 7 + 4 + 1, 7 + 3 + 2$ and $5 + 4 + 3$, and $p_{e,e}(12) = 3$ counts $10 + 2$, $8 + 4$ and $6 + 3 + 2 + 1$. Show that

$$p_{o,e}(n) - p_{e,o}(n) = \begin{cases} 1 & \text{if } n = k(3k - 1)/2 \text{ and } k \text{ is even,} \\ -1 & \text{if } n = k(3k + 1)/2 \text{ and } k \text{ is odd,} \\ 0 & \text{otherwise,} \end{cases}$$

and

$$p_{o,o}(n) - p_{e,e}(n) = \begin{cases} 1 & \text{if } n = k(3k-1)/2 \text{ and } k \text{ is odd,} \\ -1 & \text{if } n = k(3k+1)/2 \text{ and } k \text{ is even,} \\ 0 & \text{otherwise.} \end{cases}$$

76. (Periodic Cellular Automata)   Let G be a graph in which every vertex has an odd degree. Let $f_0, f_1, \ldots$ be functions mapping the vertex set $V(G)$ into $\{0, 1\}$. For $v \in V(G)$ and $t \geq 0$ we call $f_t(v)$ the *state of v at time t*; also, f_t is the *configuration* at time t.

Such a sequence $(f_t)_0^{\infty}$ is the *majority bootstrap percolation on G with initial configuration f_0* if $f_t(v)$, the state of $v \in V(G)$ at time t, is the state of the majority of the neighbours of v at time $t - 1$, i.e. $f_t(v) = \omega$ if

$$|\{u \in \Gamma(v) : f_{t-1}(u) = \omega\}| > |\{u \in \Gamma(v) : f_{t-1}(u) \neq \omega\}|.$$

Note that the entire sequence $(f_t)_0^{\infty}$ is determined by the initial configuration f_0: the update rule is the simple majority in which at each time step every vertex goes into the preferred state of its neighbours. This update happens simultaneously: the state of every vertex gets updated at every step.

A majority percolation $f = (f_t)_0^{\infty}$ is *s-periodic* if $f_{t+s} = f_t$ for every sufficiently large t. The *period* of f is the minimal s such that f is s-periodic. Since our graph G is finite, there are only finitely many configurations (at most 2^n if there are n vertices); consequently, every sequence $f = (f_t)_0^{\infty}$ is periodic with a finite period.

Trivially, f may have periods 1 and 2; e.g. if G is bipartite with bipartition $V_0 \cup V_1$ and $f_0(v) = \omega$ whenever $v \in V_\omega$, then $f_{2t} = f_0$ and $f_{2t+1} = f_1$ for every $t \geq 1$. After this very long preparation, we can state our problem.

Show that the majority percolation on a graph whose every degree is odd is periodic with period one or two: no other period is possible.

77. (Meeting Set Systems)   Let us say that a set *meets* a set system if it meets (intersects) every set in the set system. Let $\mathcal{A} = \{A_1, \ldots, A_n\} \subset S^{(\leq k)}$ be a system of n subsets of a finite set S, with each A_i having at most k elements, such that every element of S is contained in at most d of the A_i. Let X_p be a p-random subset of S, i.e. a subset obtained by selecting every integer i, $1 \leq i \leq N$, independently, with probability p. Then the probability that X_p meets $\mathcal{A}$ is at most

$$\left(1 - q^k\right)^{n/d},$$

where $q = 1 - p$.

78. (Dense Sets of Reals)   Let P be a set of points in the plane $\mathbb{R}^2$ such that the set

$$\{x/y : (x,y) \in P, y \neq 0\}$$

is dense in $\mathbb{R}$. Show that there is an $\alpha \in \mathbb{R}$ such that $\{x + \alpha y : (x,y) \in P\}$ is also dense in $\mathbb{R}$.

79. (Partitions of Boxes)   An *n-dimensional combinatorial box* or, simply, a *box*, is a set of the form $A = A_1 \times \cdots \times A_n$, where the A_i are non-empty finite sets. A *sub-box* of A is a subset of A of the form $B = B_1 \times \cdots \times B_n$; we call B *non-trivial* if $\emptyset \neq B_i \neq A_i$ for every i. [This condition is much stronger than $\emptyset \neq B \neq A$; in particular, if $|A_i| = 1$ for some i then A does not have a non-trivial sub-box.] Note that if each A_i is partitioned into two non-empty sets, B_i and C_i, then the 2^n sets of the form $D_1 \times \cdots \times D_n$, where each D_i is either B_i or C_i, give a partition of A into non-trivial boxes.

Show that the example above is best possible in the sense that an n-dimensional combinatorial box cannot be partitioned into fewer than 2^n non-trivial sub-boxes.

80. (Distinct Representatives)   Let $A_1, \ldots, A_n$ be finite sets satisfying

$$\sum_{1 \leq i < j \leq n} \frac{|A_i \cap A_j|}{|A_i||A_j|} < 1.$$

Show that the sets have distinct representatives: there are distinct elements $a_1, \ldots, a_n$ such that $a_i \in A_i$ for every i.

81. (Decomposing a Complete Graph)   A graph G is *decomposed* into its subgraphs $G_1, \ldots, G_r$ if each edge of G is an edge of precisely one G_i. [We could demand that each vertex of G belongs to at least one G_i, but that is irrelevant.] For example, a complete graph K_n with vertex set $[n]$ can be decomposed into $G_1, \ldots, G_{n-1}$, where G_i has vertex set $\{i, i + 1, \ldots, n\}$ and edge set $\{ij : i < j \leq n\}$. Note that there are many other decompositions of K_n into $n - 1$ complete bipartite graphs (see Figure 6).

Show that a complete graph on n vertices cannot be decomposed into $n - 2$ complete bipartite graphs. Equivalently, as a bipartite graph may be the trivial

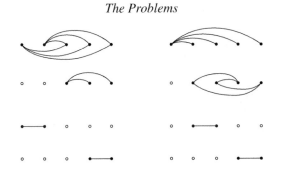

Figure 6 Two decompositions of K_5 into four complete bipartite graphs.

graph on a single vertex, a complete graph on n vertices cannot be decomposed into $n - 2$ or fewer non-trivial complete bipartite graphs.

82. (Matrices and Decompositions) (i) Check that the adjacency matrix of a graph comprising a complete bipartite graph and isolated vertices is the sum of a matrix of rank 1 and an antisymmetric matrix.

(ii) Make use of (i) to show that a complete graph on n vertices cannot be decomposed into $n - 2$ complete bipartite graphs.

83. (Patterns and Decompositions) Let the complete graph K_n with vertex set $[n]$ be decomposed into r complete bipartite graphs, $G_1, \ldots, G_r$, with G_k having bipartition (U_k, W_k). Thus U_k and W_k are disjoint subsets of $[n]$, their union is a (not necessarily proper) subset of $[n]$, and for every edge e of K_n there is precisely one k such that e joins U_k to W_k. Define the *pattern* of a map $f : [n] \to [N]$ to be the $(r + 1)$-tuple $\pi = \pi(f) = (p_1, \ldots, p_{r+1})$ with $p_k = \sum_{i \in U_k} f(i)$ for $1 \leq k \leq r$, and $p_{r+1} = \sum_{i=1}^{n} f(i)$.

Show that if $r \leq n - 2$ and N is sufficiently large then there are maps $f, g : [n] \to [N]$ with the same pattern. Deduce from this that r must be at least $n - 1$.

84. (Six Concurrent Lines) Let P_1, P_2, P_3 and P_4 be four points on a circle. For $1 \leq i < j \leq 4$ let ℓ_{ij} be the line through the midpoint of the segment $P_i P_j$ and perpendicular to the line $P_h P_k$, where $\{i, j, h, k\} = \{1, 2, 3, 4\}$, as in Figure 7. Show that the six lines ℓ_{ij} go through the same point.

85. (Short Words – First Cases) Let $A = \{a_1, \ldots, a_k\}$ be a finite set with $k \geq 2$ elements, an *alphabet*. Also, A^ℓ is the set of all *words* of *length* ℓ, and $A^* = \bigcup_\ell A^\ell$ is the set of all finite words. [This set A^* contains the *empty word* ε as well.] E.g. if $A = \{a, b, c\}$ then $A^2 = \{aa, ab, ac, ba, bb, bc, ca, cb, cc\}$. A word $x \in A^*$ is a *factor* of $w \in A^*$ if $w = uxv$ for some words $u, v \in A^*$, where uxv and similar expressions stand for concatenation. A word w *avoids* a set

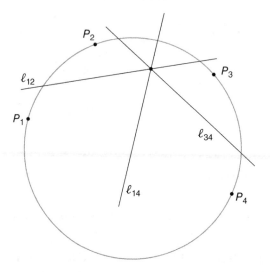

Figure 7 The lines ℓ_{12}, ℓ_{14} and ℓ_{34}.

$X \subset A^*$ if no word in X is a factor of w, and a set $X \subset A^*$ is *unavoidable* if all but finitely many words in A^* have a factor in X. For example, the set $\{a, b^4, c^5, bc\} = \{a, bbbb, ccccc, bc\}$ is unavoidable over the alphabet $\{a, b, c\}$, but the set $\{a, bc^2, b^2c\} = \{a, bcc, bbc\}$ is avoidable.

For $m \geq 1$, write $n_k(m)$ for the minimal cardinality of an unavoidable set in which each word has length at least m. Show that $n_k(1) = k$, $n_k(2) = \binom{k+1}{2}$ and $n_2(3) = 4$.

86. (Short Words – The General Case)   Let $A = \{a_1, \ldots, a_k\}$ be a finite set with $k \geq 2$ elements, an *alphabet*. Also, A^ℓ is the set of all *words* of *length* ℓ, and $A^* = \bigcup_\ell A^\ell$ is the set of all finite words. [This set A^* contains the *empty word* ε as well.] For example, if $A = \{a, b, c\}$ then $A^2 = \{aa, ab, ac, ba, bb, bc, ca, cb, cc\}$. A word $x \in A^*$ is a *factor* of $w \in A^*$ if $w = uxv$ for some words $u, v \in A^*$, where uxv and similar expressions stand for concatenation. A word w *avoids* a set $X \subset A^*$ if no word in X is a factor of w, and a set $X \subset A^*$ is *unavoidable* if all but finitely many words in A^* have a factor in X. For example, the set $\{a, b^4, c^5, bc\} = \{a, bbbb, ccccc, bc\}$ is unavoidable over the alphabet $\{a, b, c\}$, but the set $\{a, bc^2, b^2c\} = \{a, bcc, bbc\}$ is avoidable.

For $m \geq 1$, write $n_k(m)$ for the minimal cardinality of an unavoidable set in which each word has length at least m. Show that

$$k^m/m \leq n_k(m) \leq k^m.$$

87. (The Number of Divisors) Write $d(n)$ for the number of divisors of a natural number n. Thus $d(1) = 1$, $d(2) = 2$, $d(6) = 4$, $d(72) = 12$, etc. Show that $d(n) \le 2\sqrt{n}$ and $d(n) = n^{o(1)}$.

88. (Common Neighbours) Show that every graph of order n and size m has two vertices with at least $\lfloor 4m^2/n^3 \rfloor$ common neighbours.

89. (Squares in Sums) Let A be a set of n integers such that $A + A$ contains $1, 4, 9, \ldots, m^2$, the first m squares. How small can n be?

90. (Extension of Bessel's Inequality) Let $\varphi_1, \varphi_2, \ldots, \varphi_n$ and f be vectors in a real Hilbert space H. Show that

$$||f||^2 \ge \sum_{i=1}^{n} (f, \varphi_i)^2 / s_i,$$

where $(\ ,\ )$ denotes the inner product in H and $s_i = \sum_{j=1}^{n} |(\varphi_i, \varphi_j)|$.

Recall that Bessel's inequality states that if $e_1, e_2, \ldots$ is an orthonormal sequence of vectors in a Hilbert space with inner product $(\ ,\ )$, i.e. $||e_i|| = 1$ and $(e_i, e_j) = 0$ for all $i \ne j$, then $\sum_i |(f, e_i)|^2 \le ||f||^2$ for every vector f. Thus Bessel's inequality is a special case of the inequality above.

91. (Equitable Colourings) Let S be a finite set of points in the plane $\mathbb{R}^2$. Show that there is a colouring of the points of S with two colours such that every coordinate line (i.e. a line parallel with one of the axes) is *equitably* coloured, i.e. contains at most one more point of one colour than of the other.

92. (Scattered Discs) Let $D_1, \ldots, D_n$ be unit discs (in the plane) with centres $c_1, \ldots, c_n$, $n \ge 3$, such that no line meets more than two of them. Show that

$$\sum_{1 \le i < j \le n} \frac{1}{d_{ij}} < \frac{n\pi}{4},$$

where $d_{ij} = d(c_i, c_j)$ is the distance between c_i and c_j.

93. (East Model) A beautiful model of statistical physics, the *East Model*, is defined as follows. The board (our universe) is the half-line formed by the natural numbers $1, 2, \ldots$: we call them *sites*. Each site is in one of two *states*: it is either *occupied* or *unoccupied*. We call the set of occupied sites a *configuration*. (Equivalently, we may consider the distribution of the states as a *configuration*.) An *East process* or simply *process* is a sequence of configurations $X_0 \to X_1 \to \cdots \to X_\ell$ such that X_{j+1} is obtained from X_j by changing the state of at most one of the sites, and the state of a site x is allowed to change only if $x = 1$ or $x - 1$ is occupied.

Denote by $V(n)$ the set of configurations that can be created by these processes if we start without any occupied sites and at each step there are at most n occupied sites. (As a sanity check, note that $V(1) = \{\emptyset, \{1\}\}$ and $V(2) = \{\emptyset, \{1\}, \{1,2\}, \{2\}, \{2,3\}\}$, so $|V(1)| = 2$ and $|V(2)| = 5$.) Determine the functions

$$A(n) = \max\{x : \{x\} \in V(n)\}$$

and

$$B(n) = \max\{x : x \in X \text{ for some } X \in V(n)\}.$$

Thus $A(n)$ is the greatest value of x for which the configuration in which only x is occupied belongs to $V(n)$, and $B(n)$ tells us 'how far out' we can reach while in $V(n)$. The examples above tell us that $A(1) = B(1) = 1$, $A(2) = 2$ and $B(2) = 3$.

94. (Perfect Triangles)   In this problem we call a triangle *perfect* if it has integer sides and its area is equal to its perimeter length. Determine all perfect triangles.

95. (A Triangle Inequality)   Let a, b and c be the sides, and Δ the area of a triangle. Show that

$$a^2 + b^2 + c^2 \geq 4\sqrt{3}\Delta.$$

96. (An Inequality for Two Triangles)   We are given two triangles: one with sides a, b, c and area Δ, and another with sides a', b', c' and area Δ'. Show that

$$a'^2(-a^2 + b^2 + c^2) + b'^2(a^2 - b^2 + c^2) + c'^2(a^2 + b^2 - c^2) \geq 16\Delta\Delta'.$$

97. (Random Intersections)   For $0 < p < 1$, a *p-random subset* of $[n] = \{1, 2, \ldots, n\}$ is obtained by picking each number k, $1 \leq k \leq n$, with probability p, independently of all other choices. Thus the probability that we pick a set $A \subset [n]$ is

$$\mathbb{P}_p(A) = \mathbb{P}(X_p = A) = p^{|A|}(1-p)^{n-|A|}.$$

Equivalently, the probability measure $\mathbb{P}_p$ on $\mathcal{P}_n$, the set of all 2^n subsets of $[n]$, is given by

$$\mathbb{P}_p(\mathcal{A}) = \sum_{A \in \mathcal{A}} \mathbb{P}_p(A) = \sum_{A \in \mathcal{A}} p^{|A|}(1-p)^{n-|A|},$$

where $\mathcal{A} \subset \mathcal{P}_n$. After this preparation, we can state our problem.

Let $\mathcal{A} \subset \mathcal{P}_n$ and set

$$J = J(\mathcal{A}) = \{A \cap B : A, B \in \mathcal{A}\}.$$

Show that for $0 < p < 1$ we have

$$\mathbb{P}_{p^2}(\mathcal{J}) \geq \mathbb{P}_p(\mathcal{A})^2.$$

98. (Disjoint Squares)   A *standard square* in $\mathbb{R}^2$ is a square of the form $\{(x, y) \in \mathbb{R}^2 : u \leq x \leq u + s, \, w \leq y \leq w + s\}$, where $s > 0$ is the side length. (Thus we take a standard square closed, with the sides parallel to the axes.) Let $\mathcal{F} = \{Q_1, \ldots, Q_n\}$ be a family of standard unit squares in $\mathbb{R}^2$ with their union $A = \bigcup_{i=1}^n Q_i$ having area greater than $4k$. Show that $\mathcal{F}$ contains a subcollection of $k + 1$ pairwise-disjoint unit squares.

Show also that if the area of A is $4k$, we cannot guarantee $k + 1$ disjoint squares.

99. (Increasing Subsequences)   (i) Show that every sequence of $pq + 1$ real numbers contains either a strictly increasing subsequence of length $p + 1$ or a decreasing subsequence of length $q+1$. To spell it out, if $a_1, \ldots, a_n$ is a sequence of $n = pq + 1$ real numbers, then there is a subsequence $a_{i_0} < a_{i_1} < \cdots < a_{i_p}$ or a subsequence $a_{j_0} \geq a_{j_1} \geq \cdots \geq a_{j_q}$.

(ii) Show also that the assertion in (i) is best possible, i.e. $pq + 1$ cannot be replaced by pq.

100. (A Permutation Game)   A teacher plays the *Permutation Game* with her students. She sticks a piece of paper on the forehead of each of the six best students of the Maths Club, and writes a (real) number on each, making sure that no two are the same. She challenges the students to divide themselves into two groups, group A and group B, with one consisting of the students with the largest, third largest and fifth largest numbers, and the other consisting of the remaining three students. Only the division is required: group A may contain the largest number or the smallest. Needless to say, no student is allowed to get *any* help from the others, and they have to declare simultaneously whether they join group A or group B; however, they are allowed to decide on their strategy. What strategy can the students use to win the Permutation Game?

101. (Ants on a Rod)   There are 50 ants on a rod of length 1 m. Each ant is hurrying along at $10\,\text{mm/sec}$ in a fixed direction. When two ants meet, they turn around and each hurries in the opposite direction with the same speed. An ant that reaches the end of the rod falls off. At most how long will it take for all the ants to fall off?

102. (Two Cyclists and a Swallow)   Adalbert starts to cycle from town A towards town B, which is 60 km away, doing 20 km/h. At the same time, his girl-friend, Bernadette, cycles from town B towards town A, doing 10 km/h. Their pet swallow, Fouché, doing 40 km/h, starts with Adalbert from A towards B:

when it reaches Bernadette, it turns round and flies back to Adalbert, etc. What distance will Fouché cover by the time Adalbert and Bernadette meet?

103. (Almost Disjoint Subsets of Natural Numbers)   As in Problem 10 of *Coffee Time in Memphis*, we call two sets *almost disjoint* if their intersection is finite. We know from that problem that the countable set of natural numbers has an uncountable family of almost disjoint subsets. Is there an uncountable family $\{M_\gamma : \gamma \in \Gamma\}$ of almost disjoint sets of natural numbers such that every pairwise intersection $M_\alpha \cap M_\beta$ is an initial segment of both M_α and M_β?

104. (Primitive Sequences)   We call a sequence $(a_1, \ldots, a_n)$ of real numbers *primitive* if $|ka_i - a_j| \geq 1$ whenever $1 \leq i, j \leq n$, $i \neq j$, and k is an integer. Given $b \geq 1$, what is the maximal value of n for which there is a primitive sequence $(a_1, \ldots, a_n)$ with $0 < a_i \leq b$ for every i?

105. (The Time of Infection on a Grid)   Let G_n be an $n \times n$ board or grid made up of n^2 unit squares, *cells*. Equivalently, G_n is an $n \times n$ part of the lattice $\mathbb{Z}^n$, made up of n^2 vertices or *sites*, and $2n(n-1)$ edges connecting them. We shall switch freely between these two ways of viewing G_n, see Figure 8.

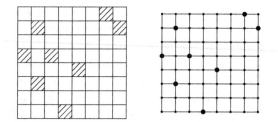

Figure 8 Two representations of G_8, with eight initially infected cells/sites leading to full infection.

At time 0, we have a set S_0 of *initially infected* sites. At each time step, every site with at least two infected neighbours gets infected, while the sites that have already been infected, remain infected for ever. The process stops in the tth step, at time t, if after that no new sites will be infected. Formally, for $t \geq 1$, put

$$S_t = \{x \in S \setminus S_{t-1} : x \text{ has at least two neighbours in } S_{t-1}\} \cap S_{t-1}.$$

We say that S_0 *infects the entire grid* G_n if, for some time t, the set S_t is the entire vertex set $V(G_n)$. The minimal t with this property is the *infection time* of S_0. We know from Problem 34 of *Coffee Time in Memphis* that the minimal number of sites needed to infect the entire grid G_n is n. What is the minimal time of full infection if we start with a set of n initially infected sites?

The reader may care to check that the seven initially infected cells in Figure 9 take fourteen steps to infect G_7.

10	9	8	7		7	8
	8	7	6	5	6	7
10	9	4		4	5	6
11	10	3	2	3	4	
12	11		1		5	6
13	12	9	8	7	6	7
14	13	10	9	8		8

Figure 9 The numbering showing how seven initially infected cells infect G_7 in 14 steps.

106. (Areas of Triangles)   (i) Given a triangle ABC, let D, E and F be points on the sides AB, BC and CA, respectively, as in Figure 10(i). Show that the segments AD, BE and CF are *Cevians*, i.e. are concurrent, if and only if

$$\frac{AF}{FB} \cdot \frac{BD}{DC} \cdot \frac{CE}{EA} = 1.$$

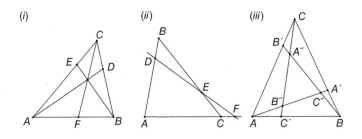

Figure 10 Points, lines and triangles.

(ii) A line meets the sides AB and BC of a triangle ABC in D and F, and the extension of the side AC in F, as in Figure 10(ii). Show that

$$\frac{AD}{DB} \cdot \frac{BE}{EC} \cdot \frac{CF}{FA} = 1.$$

(iii) On the sides BC, CA, AB of a triangle, three points A', B', C' are taken such that

$$BA' : A'C = p_1 : q_1, \qquad CB' : B'A = p_2 : q_2, \qquad AC' : C'B = p_3 : q_3.$$

The segments BB' and CC', CC' and AA', AA' and BB' intersect in A'', B'', C'', as in Figure 10(iii). Show that the area of the triangle $A''B''C''$ is to the area of the triangle ABC as

$$(p_1p_2p_3-q_1q_2q_3)^2 : (p_2p_3+q_2q_3+p_2q_3)(p_3p_1+q_3q_1+p_3q_1)(p_1p_2+q_1q_2+p_1q_2).$$

107. (Lines and Vectors)   (i) Let O be the circumcentre of a triangle ABC, M the intersection of the medians, and H the orthocentre, i.e. the intersection of the altitudes (heights). Show that O, M and H are collinear. Show also that $\overrightarrow{MH} = 2\overrightarrow{OM}$.
   (ii) What is the resultant of the vectors $\overrightarrow{OA}$, $\overrightarrow{OB}$ and $\overrightarrow{OC}$?

108. (Feuerbach's Remarkable Circle)   Let H be the intersection of the altitudes AD, BE and CF of a triangle ABC. Show that the midpoints of the sides, the midpoints of the segments AH, BH, CH, and the feet D, E, F of the altitudes are on a circle.

109. (Ratio–Product–Sum Problem)   Let ABC be a triangle, and x, y, z positive numbers. Show that the following two assertions are equivalent:
(i) There are points X, Y and Z on the sides BC, CA and AB such that the

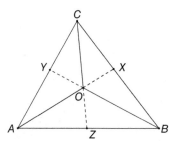

Figure 11 The concurrent segments.

segments AX, BY and CZ are concurrent in a point O (see Figure 11), and $AO/OX = x$, $BO/OY = y$ and $CO/OZ = z$.
(ii) The numbers x, y and z satisfy

$$xyz = x + y + z + 2.$$

110. (Bachet's Weight Problem)

There is a weighing scale with two pans and n integral weights. For example if our weights are 1, 2 and 8 then we can weigh all integer weights up to 11 by putting weights in either pan. What is the maximal integer W_n such that any  integral number of pounds up to W_n can be weighed when weights may be put in either pan? Determine the set of weights achieving the maximum.

111. (Perfect Partitions)   We have a weighing scale with two pans. One of the pans is for our integral weights, the other is for the object we wish to weigh.

In how many ways can you partition a weight of 31 lb into integral weights so as to be able to weigh, in only one manner, any weight of an integral number of pounds from 1 to 31 inclusive? (Weights of the same weight are considered to be identical.) For example, $(8, 8, 8, 2, 2, 2, 1)$ is such a partition: with three weights of 8 lb, three weights of 2 lb and one weight of 1 lb we may weigh any of the weights $1, 2, \ldots, 31$ in precisely one manner. (As this example shows, weights of the same weight are considered to be identical.) For example, 21 lb is weighed as follows: $21 = 8 + 8 + 2 + 2 + 1$.

112. (Countably Many Players)   Countably many players, $P_1, P_2, \ldots$, are lined up in a row; each of them has a real number on his head, which can be seen only by the players preceding him. Thus player P_3 has no idea what numbers are on P_1, P_2 and P_3 (himself), but can see the numbers on P_4, P_5, etc. The task of the players is to guess their own number, and write it on a piece of paper that no other player will see. Before the numbers are distributed, the players are allowed to agree on a strategy, but once the numbers have been placed on the heads, no communication is permitted. Can the players achieve that all but finitely many of them guess their numbers correctly?

113. (One Hundred Players)   This game is played by one hundred players; as usual, the players are allowed to agree on a strategy before the game starts, but after the start no consultation is permitted. One by one, the players are shown into a room with an infinite sequence of drawers, $D_1, D_2, \ldots$, with drawer D_n containing a real number x_n not known to any of the players. Each player is allowed to open as many drawers as he likes, but at some stage he has to point

at an unopened drawer, and guess the real number in it. It is up to him how he

decides which drawers he opens: he may open some, then think, calculate, toss a coin ten times, think again, and open some more drawers, etc. But he cannot postpone the inevitable for ever: he has to guess the content of his favourite drawer. Having made his selection and guess, the drawers are closed again, and the next player enters the room. Can the players guarantee that at least one of them guesses correctly? Maybe more than one?

114. (**River Crossings – Take One**)   (i) *De viro et muliere ponderantibus plaustrum* [A very heavy man and woman]. A man and a woman, each the weight of a cartload, with two children who together weigh as much as a cartload, have to cross a river. They find a boat which can only take one cartload. Make the transfer if you can, without sinking the boat.

(ii) *De lupo et capra et fasciculo cauli* [A wolf, a goat, and a bunch of cabbages]. A man had to take a wolf, a goat, and a bunch of cabbages across a river. The only boat he could find could only take two of them at a time. But he had been ordered to transfer all of these to the other side in good condition. How could this be done?

115. (**River Crossings – Take Two**)
De tribus fratribus singulas habentibus sorores [Three friends and their sisters]. Three men, each with a sister, needed to cross a river. Each one of them coveted the sister of another. At the river, they found only a small boat, in which only two of them could cross at a time. How did they cross the river, without any of the women being defiled by the men?

116. (Fibonacci and a Medieval Mathematics Tournament)   Emperor Frederick II (1194–1250), a grandson of Barbarossa, stopped at Pisa in 1225 in order to hold a mathematics tournament to test the skill of *Leonardo Fibonacci* (called *Leonardo of Pisa* or just *Leonardo* in his life time). The competitors were informed beforehand of the questions to be asked. Here are two of the questions.

(i) Find a number of which the square, when either increased or decreased by 5, would remain a square. The tacit assumption is, of course, that the number we have to find is rational.

(ii) Three men, A, B, C, possess a sum of money u, their shares being in the ratio $3 : 2 : 1$. A takes away x, keeps half of it, and deposits the remainder with D; B takes away y, keeps two-thirds of it, and deposits the remainder with D; C takes away all that is left, z, keeps five-sixths of it, and deposits the remainder with D. This deposit with D is found to belong to A, B and C in equal proportions. Find u, x, y and z.

Fibonacci solved both problems, but the other competitors could not solve either. Give your solutions.

117. (Triangles and Quadrilaterals)   (i) Let ABC be a triangle such that the foot D of its altitude from A is on the side BC. Suppose that, with the obvious notation, $AC - AB = 3$, $DC - DB = 12$ and $AD = 10$. Determine the length of the base BC.

(ii) Suppose that there are quadrilaterals with side lengths a, b, c and d in a cyclic order. Construct the one which is inscribed in a circle.

118. (The Cross-Ratios of Points and Lines)   Given four collinear points, A, B, C and D, define the *cross-ratio* $[A, B; C, D]$ as

$$[A, B; C, D] = \frac{AC}{AD} \bigg/ \frac{BC}{BD} = \frac{AC \cdot BD}{AD \cdot BC}.$$

Here, for three collinear points P, Q and R, we write PQ/PR for the signed ratio of the distances PQ and PR: this ratio has sign -1 if P is between Q and R, and sign $+1$ otherwise. For example, for the points $A = (0,0)$, $B = (1,0)$, $C = (2,0)$ and $D = (3,0)$ on the x-axis, $[A, B; C, D] = (2/3) \cdot 2 = 4/3$, $[A, C; B, D] = (2/3) \cdot (-1) = -1/3$ and $[B, D; C, A] = (-1) \cdot 3 = -3$.

Similarly, given four concurrent lines, a, b, c and d, define the *cross-ratio* $[a, b; c, d]$ as

$$[a, b; c, d] = \frac{\sin(ac)}{\sin(ad)} \bigg/ \frac{\sin(bc)}{\sin(bd)} = \frac{\sin(ac) \cdot \sin(bd)}{\sin(ad) \cdot \sin(bc)}.$$

To get these ratios, orient the lines a, b, c and d in any way, and define (ac) to be the angle needed to rotate a to c, etc. It is immediate that $[a, b; c, d]$ is independent of the orientation we have chosen. For example, if $(ab) = (bc) = (cd) = \pi/6$, so that $(ad) = \pi/2$, then $[a,b;c,d] = (\sqrt{3}/2) \cdot (\sqrt{3}) = 3/2$ $[a,c;b,d] = (1/2) \cdot (-1) = -1/2$ and $[b,d;c,a] = (-1) \cdot (2) = -2$.

Finally, let A, B, C and D be points on a line ℓ, and let O be a fifth point, not on ℓ. Define

$$O[ABCD] = [a,b;c,d],$$

where a is the line through O and A, b is the line through O and B, etc.

After these definitions, we can state our simple problems concerning the basic properties of the cross-ratio of points and lines.

(i) Let A, B, C and D be four points on a line ℓ, and let O be a fifth point not on ℓ. Show that

$$O[ABCD] = [A,B;C,D].$$

(ii) Let A, B, C, D, O and O' be six points on a circle. Show that

$$O[ABCD] = O'[ABCD].$$

(iii) Let $[A,B;C,D] = [A',B;C,D]$, where the two sets of arguments are collinear, and A, C, D and A', C, D are in the same order. Show that $A = A'$. Furthermore, show the analogous assertion for lines and their cross-ratios.

119. (Hexagons in Circles – Take One)   Let A, B, C, D, E and F be points on a conic section (i.e. ellipse, parabola or hyperbola) and suppose that the opposite sides of the 'hexagon' $ABCDEF$ meet in G, H and I, as in Figure 12. Thus the lines AB and DE meet in G, BC and EF meet in H, and CD and FA

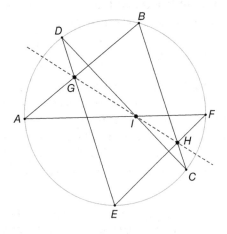

Figure 12 A hexagon inscribed in a circle, and the intersections of the opposite sides.

meet in I. Show that the points G, H and I are on a line.

120. (Hexagons in Circles – Take Two)   Let the vertices of a hexagon lie on a circle, with the three pairs of opposite sides intersecting. Show that the points of intersection are collinear.

121. (A Sequence in $\mathbb{Z}_p$)   Let p be a prime, and let $a_1, \ldots, a_{p-1}$ be a sequence of elements of $\mathbb{Z}_p$, the integers modulo p, such that $\sum_{i \in I} a_i \neq 0$ whenever I is a non-empty set of indices. Show that this sequence is constant: $a_1 = \cdots = a_{p-1}$.

122. (Elements of Prime Order)   Let G be a finite group whose order is divisible with a prime p. Show that one plus the number of elements of order p in G is divisible by p.

123. (Flat Triangulations)   At most how many triangles are there in a triangulation of an n-gon in which every internal vertex has degree 6?

124. (Triangular Billiard Tables)   Let ABC be a triangular billiard table with acute angles. For which points P, Q and R on the sides BC, CA and AB is it true that a billiard ball, launched from P towards Q, describes the 3-periodic polygonal path $PQRPQR\cdots$? Every 'bounce' is taken to be 'true': the ball

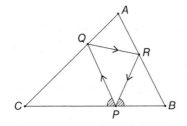

Figure 13 The trajectory of a billiard ball on a triangular table.

leaves a side at exactly the same angle at which it impacts it. Thus RP and QP form equal angles with the side BC. (See Figure 13.)

125. (Chords of an Ellipse)   Let AB be a chord of an ellipse with midpoint M, and let PQ and RS be two other chords through M. Denote by T and U the intersections of the chords PS and RQ with AB, as in Figure 14. Show that M is the midpoint of the segment TU.

126. (Recurrence Relations for the Partition Function)   Recall that we write $p(n)$ for the *partition function*, the number of unordered partitions of n into non-negative integers, so that $p(1) = 1$, $p(2) = 2$, $p(3) = 3$, $p(4) = 5$, $p(5) = 7$, etc. Also, we define $p(0) = 0$: after all, 0 has just one partition, the empty partition, the partition that has no parts. Furthermore, $p(-m) = 0$ for

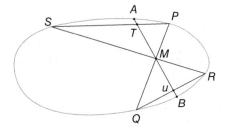

Figure 14 The butterfly with wings PMS and RMQ.

every $m \geq 1$.

Show that for $n \geq 1$ we have

$$np(n) = \sum_{k=1} \sum_{v=1} vp(n - kv)$$

and

$$p(n) = p(n-1) + p(n-2) - p(n-5) - p(n-7) + p(n-12) \pm \cdots$$
$$= \sum_{k=1} (-1)^{k+1} \left(p\left(n - \frac{k(3k-1)}{2}\right) + p\left(n - \frac{k(3k+1)}{2}\right) \right).$$

In the summations above, k takes all the positive values that do not give negative arguments for the partition function; thus, in the second summation k takes about $\sqrt{2n/3}$ values.

127. (The Growth of the Partition Function)   As before, let $p(n)$ be the partition function, the number of partitions of n, so that $p(0) = 1$, $p(1) = 1$ and, starting with $p(2)$, the values of $p(n)$ are as follows: 2, 3, 5, 7, 11, 15, 22, 30, 42, 56, 77, 101, ... Show that

$$p(n) \leq e^{c\sqrt{n}}$$

where $c = \pi\sqrt{2/3} = 2.565 \cdots$.

In your argument you may need the following inequality: if $0 < x < 1$ then

$$e^{-x}/(1 - e^{-x})^2 < 1/x^2.$$

128. (Dense Orbits)   Show that there is a bounded linear operator $T \in \mathcal{B}(\ell^1)$ such that for some vector $x \in \ell^1$ the orbit $\{T^n x : n = 1, 2, \ldots\}$ is dense in ℓ^1. Here ℓ^1 is the classical sequence space

$$\left\{ x = (x_i)_1^\infty : x_i \in \mathbb{R}, \ ||x|| = \sum_{i=1}^{\infty} |x_i| < \infty \right\}.$$

The Hints

2. Be greedy: having found $2 \le n_1 < n_2 < \cdots < n_i$, if

$$r > \frac{1}{n_1} + \cdots + \frac{1}{n_i}$$

then choose $1/n_{i+1}$ as large as possible subject to the condition

$$r \ge \frac{1}{n_1} + \cdots + \frac{1}{n_{i+1}}.$$

Check that this works for $\frac{4699}{7320}$; learning from this example, prove that it works for every rational number.

3. Make use of Sylvester's result from Problem 2, 'Vulgar Fractions'.

5. Let X_p and Y_p be independent p-random subsets of $[n]$. Note that $X_p \cap Y_p$ is a p^2-random subset of $[n]$.

6. To this end, start with the power series expansion of $\sin x$ that can be seen from $\sin x = (e^{-ix} - e^{ix})$, and consider the arising power series expansion of $(\sin x)/x$.

7. In the first proof, give upper and lower bounds for the logarithm of the product $\prod_{p \le n} (1 - 1/p)^{-1}$: the upper bound in terms of $1/p$ and the lower bound in terms of the logarithm of $\sum_{k=1}^{n} 1/k$.

In your second and third proofs, let $p_1 = 2 < p_2 = 3 < \cdots$ be the sequence of primes. Assuming that $\sum_1^{\infty} 1/p_i < \infty$, let k be such that $\sum_{i=k+1}^{\infty} 1/p_i$ is 'small', less than $1/2$ or $1/8$, or whatever your choice is. Then, for n large, partition the integers up to n into two parts: those that have a prime factor p_ℓ with $\ell > k$ and those that do not. Give two proofs of the fact that these two parts are too small for their union, being $[n] = \{1, \ldots, n\}$.

33

9.   To prove that the conditions are sufficient, add a row to our $r \times n$ matrix in such a way that the $(r + 1) \times n$ matrix obtained satisfies the conditions, i.e. each of its $r + 1$ rows has precisely k ones, and each of its n columns has at least $k + r + 1 - n$ and at most k ones.

10.   Use an argument from physics.

11.   Show that the average number of faces meeting a face is more than five. Make use of Euler's polyhedron theorem and the fact that if the vertices of a face have degrees $d_1, \ldots, d_\ell$ then this face meets $\sum_i (d_i - 2)$ other faces.

13.   Suppose that the assertion is false. Thus, let ABC be a triangle with angles α and β at A and B, with $\alpha < \beta$, and angle bisectors AD and BE of the same length, with D on the side BC and E on AC. Let F be the point on the bisector AD such that the angle ABF is $(\alpha + \beta)/2$ and so the angle EBF is $\alpha/2$.

14.   Add a point F on the side AC such that $\angle CBF = 20°$.

15.   To prove that the angle is $20°$, reflect A in the line determined by CD to get a point H. Chase angles.

16.   For an algebraic proof, make use of the identity $b^2 = (c + a)(c - a)$. For a geometric proof, consider the line in the plane through the points $(-1, 0)$ and $(a/c, b/c)$.

17.   Prove more, namely that $a^4 + b^4 = c^2$ has no solutions in positive integers. To this end, following your nose, repeatedly make use of the characterization of Pythagorean triples in the preceding problem. As a start, note that if (a, b, c) is a solution then (a^2, b^2, c) is a Pythagorean triple.

18.   Clearly, 1 is a congruent number if there is a right-angled triangle with integer sides whose area is a square. Supposing that there is such a triangle, take one whose hypotenuse is as small as possible, and construct another right-angled triangle with integer sides whose area is a square and whose hypotenuse is even smaller. To carry out your construction, recall the characterization of Pythagorean triples.

19.   The main problem is to find necessary conditions: then it is easy to check that these conditions are also sufficient. Assuming that $\sqrt{s + 1} - \sqrt{s - 1} = a/b$ for relatively prime numbers a and b, square this equation and draw the appropriate divisibility consequences and recall the characterization of Pythagorean triples.

21. Let P_n be an n-gon of perimeter length n with maximum area. (Although the existence of such a polygon P_n is not obvious, in this solution we assume it.) First, note that P_n is equilateral – this is very easy; this implies our result for $n = 3$ and $n = 4$. Second, assuming that $n \geq 5$ and P_n is not equiangular, let $ABCD$ be four consecutive vertices, with the angle at B not equal to the angle at C. Now reflect $ABCD$ in the perpendicular bisector of AD, to obtain $AC'B'D$. Check that the 'arithmetic mean' of the two congruent quadrilaterals $ABCD$ and $AC'B'D$ is a quadrilateral of smaller perimeter and larger area.

22. A hint would make the solution too easy.

23. Use the previous problem to deduce your first proof.

25. Let $ABCD$ be a convex quadrilateral, with E and F the midpoints of the diagonals AC and BD. What is the locus of the points O of the quadrilateral such that $AOB + COD = BOC + DOA$, where XYZ denotes not only a triangle, but also the area of that triangle?

To answer this question, argue with the aid of the areas of various triangles.

26. In your first proof of (ii) use generating functions, in the second give a slick concatenation argument for a one-to-one correspondence, and in the third classify the partitions according to the multiplicity of 1 in the partitions.

27. Use generating functions.

28. I have no doubt that every reader has noticed that (ii) is more general (i.e. trivially contains) part (i). For once, the hope is not that the simpler part is done first: it is very likely that, using generating functions, in (ii) one is more likely to be led to a simple proof.

29. Deduce from the Sylvester representation of a rational number as a sum of reciprocals that for every positive rational r and every positive integer A there is a finite sequence $(n_i)_1^\ell$ of natural numbers such that

$$A < n_1 < n_1 + A < n_2 < n_2 + A < n_3 < \cdots < n_{\ell-1} + A < n_\ell$$

and

$$r = \sum_1^\ell \frac{1}{n_i}.$$

30. Note that our conditions are that $\sum_{h=1}^n f_i(h) = r$ for every i, and

$$\sum_{h=1}^n f_i(h)f_j(h) \leq s < r^2/n$$

whenever $i \neq j$. Bound $\sum_{h=1}^{n} F(h)^2$ from above and below to deduce the required inequality.

31. Consider the rows and columns as 0, 1 sequences of length n, and note that no matter in what permutation you take the rows, the new diagonal 'intersects' every row.

34. The answer is $(n!)^3$. Prove it in a very simple way, without the use of generating functions.

35. Prove that every pair of points of S is on a halving circle. Given points p and q, find a circle through p and q that contains at least n points in its interior, and move it continuously, keeping p and q on it, towards a position with at most $n - 1$ points in its interior.

36. Construct the set of $2n + 1$ points and the union of two sets, one with $n + 1$ points and the other with n. The sets should be such that every halving circle has two points in one and one in the other.

37. For which polynomials $f(X)$ can we prove in one step that the linear combination is zero?

38. The sum is $n!$.

39. Apply induction on a, starting with the trivial case $a = 0$. To prove the induction step, introduce three functions: let $S(a)$ be the left-hand side, $F(a, k)$ the kth term on the left, and set

$$G(a,k) = \frac{(-1)^k (a + b)!(b + c)!(c + a)!}{2(a + k + 1)!(a - k)!(b + k)!(b - k - 1)!(c + k)!(c - k - 1)!}.$$

To prove that $S(a + 1) = S(a)(a + 1 + b + c)/(a + 1)$, the relation needed to prove the induction step, write $G(a, k) - G(a, k - 1)$ as an appropriate linear combination of $F(a + 1, k)$ and $F(a, k)$.

40. To prove the first inequality, show that the polynomials on the two sides of the identity agree at $X = -m - n, -m - n + 1, \ldots, n - 1, n$; in fact, they are 0 at all but n. Since the polynomials have degree at most $m + 2n$ and agree at $m + 2n + 1$ places, they are identical.

41. As in Figure 15, let $O = (0,0)$, $X_n = (1, x_n)$, $n \geq 0$, $Y_\infty = (0, 1)$, and denote by C the unit circle with centre O. For $n \geq 1$, let Y_n be the intersection of the segment OX_n with the circle C, and Z_{n-1} the intersection of the line through Y_n parallel with $X_0 X_n$ with the segment OX_{n-1}. For $n \geq 1$, denote by $|T_n|$ the area of the triangle $OY_n Z_{n-1}$, and note that $\sum_{n=1}^{\infty} |T_n| < \pi/4$.

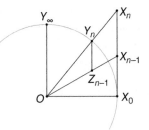

Figure 15

42. Make use of the final inequality appearing in the solution of Problem 41.

43. Apply induction on n.

44. Write out α_p in a pedestrian way, as an expression involving the terms $\lfloor 2n/p^k \rfloor$ and $\lfloor n/p^k \rfloor$ for $k = 1, 2, \ldots$ satisfying $p^k \leq 2n$.

45. To make sure that we understand what $\Pi(n)$ stands for, note that if $x \geq 2$ is a real number then $\Pi(x) = \Pi(\lfloor x \rfloor)$. Also, writing $p_1 = 2 < p_2 = 3 < \cdots$ for the sequence of primes, if $p_k \leq x < p_{k+1}$ then $\Pi(x) = \Pi(p_k)$. In particular, if $m \geq 2$ is a natural number then $\Pi(2m) = \Pi(2m-1)$. More importantly, observe that

$$\Pi(2m - 1)/\Pi(m) \leq \binom{2m - 1}{m}.$$

Recalling from an earlier problem an upper bound on the central binomial coefficient, use this ineqality to prove the results by induction on n.

46. First, show that Bertrand's postulate is easily checked up to a pretty large number, certainly up to $2,500$. Supposing that the conjecture fails for some $n \geq 2,500$ so that $\Pi(2n) = \Pi(n)$, use the results in Problems 42, 44 and 45 to give lower and upper bounds for the central binomial coefficient $\binom{2n}{n}$, and then show that these bounds are incompatible.

47. Take the powers of 2 and 3 modulo 8.

48. Consider the cases n even and n odd separately. In the case n even, write r as $r = 1 + 2^k q$, where q is odd, and consider the equation to be solved modulo 2^{k+1}.

49. Imitate the proof concerning the equation $2^m = r^n - 1$.

50. Let $p^m = r^n - 1$, where $p \geq 3$ is a prime and m, r and n are integers greater than 1. We know from Problem 49 that $r \neq 2$, so $r \geq 3$. Note that $r^n - 1 = (r - 1)(1 + r + r^2 + \cdots + r^{n-1})$, so both factors are perfect powers of p. Draw the appropriate conclusions.

53. Show that it suffices to prove the assertion in the case when the weights are 0 and 1.

56. Note that we may assume that f and g have the same degree. Assuming that S_g is a (not necessarily proper) subset of S_f, consider the ratio of our polynomials on the unbounded component of the complement of S_f.

57. Define two random variables, T and W, as follows. Write T for the number of tosses the game takes, and set

$$W = \begin{cases} 1 & \text{if Rosencrantz wins,} \\ 0 & \text{if Guildenstern wins.} \end{cases}$$

Although our task is to show that

$$\mathbb{E}(T|W = 1) = \mathbb{E}(T|W = 0),$$

let us prove more, namely that the random variables T and W are independent.

First, show that T and W are independent if for some constant $c > 0$ we have

$$\mathbb{P}(W = 1 \text{ and } T = t) = c\mathbb{P}(T = t).$$

Second, set up a one-to-one correspondence between the set $R(t)$ of strings of tosses $HTTHTHH \ldots$ resulting in a win for Rosencrantz after t tosses, and the analogous set $G(t)$ for Guildenstern. Show that

$$\mathbb{P}(R(t)) = (p/q)^k \, \mathbb{P}(G(t)),$$

and proceed to prove the required assertion.

58. If in doubt, argue the most pedestrian way.

59. Keep a cool head – the problem is trivial. If you want a pedestrian (and lengthy) solution, write out the entire decision tree.

61. (i) Consider the position of the sofa well before it reaches the corner, and attach the infinite strip corresponding to the passage to the sofa. Repeat this when the sofa gets far from the corner. What can one say about these strips and the sofa as the sofa goes round the corner?

(ii) Think of an old-fashioned telephone set.

63. Assume that q is not a perfect square, fix a, say, and consider the quadratic equation $x^2 - aqx + (a^2 - q) = 0$, one of whose roots is b. Having made a legitimate assumption about the pair (a, b), like $a + b$ is minimal, a is minimal, $a^2 + b^2$ is minimal, or whatever, show that the root of this quadratic equation different from b can play the role of b (with a, of course), and leads to a contradiction.

65. Writing $a_i = 2^{\alpha_i} b_i$, where α_i is a non-negative integer and b_i is odd, as no a_i divides another, the b_i factors are distinct. As we have n factors b_i, their sequence $b_1, b_2, \ldots, b_n$ is a permutation of $1, 3, 5, \ldots, 2n - 1$.

68. Parity.

70. Prove the analogous assertion for 'nice' measures in the plane, with the measure of a plane set corresponding to the number of our $6k$ points in the set.

72. For (ii), try polynomials of the form $1 + aX^2Y^2 + bX^4Y^2 + cX^2Y^4$.

73. Represent a partition by its *Ferrers diagram* or *Young tableau*, as in Figure 16. The number of dots or stars in the kth row of the Ferrers diagram (and the number of cells in the kth row of the Young tableau) of a partition is

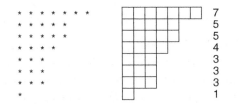

Figure 16 Two representations of the partition $31 = 7 + 5 + 5 + 4 + 3 + 3 + 3 + 1$.

the kth part of the partition.

74. Rearrange a partition of one kind to give one of the other kind, which is defined on almost all permutations into unequal parts.

75. Consider the involution Franklin constructed in his proof of Euler's Pentagonal Theorem (see Problem 74).

76. Suppose that the assertion is false for a graph G, so there is a majority bootstrap percolation $f = (f_t)_0^\infty$ on G whose period is $k \geq 3$. We may assume that the periodicity starts from the very beginning: $f_{t+k} = f_t$ for every $t \geq 0$. To reduce clutter, write v_t for the state $f_t(v)$ of a vertex v at time t, taking the suffices modulo k, so that $v_{k+1} = v_1$, etc. As the period is $k \geq 3$, there is a vertex v such that $v_t \neq v_{t-2}$ for infinitely many times t. Show that for such a pair (v,t) we have

$$\sum_{u \in \Gamma(v)} \left(\delta(u_{t-1}, v_t) - \delta(u_{t-1}, v_{t-2})\right) \geq 0,$$

where δ is the Kronecker δ function, so that $\delta(x, y) = 1$ if $x = y$ and it is 0 otherwise.

Use this to deduce that

$$\sum_{t=1}^{k} \sum_{v \in V} \sum_{u \in \Gamma(v)} \left(\delta(u_{t-1}, v_t) - \delta(u_{t-1}, v_{t-2})\right) > 0.$$

Evaluate this triple sum in two different ways to get a contradiction.

77. We may assume that $A_1 = [k] = \{1, \ldots, k\}$. For $1 \leq \ell \leq k$, let E_ℓ be the event that ℓ is the minimal element of $X_p \cap A_1 \cap$, and set

$$\mathcal{A}_\ell = \{A_i \setminus [\ell - 1] : \ell \notin A_i\}.$$

Note that X_p meets $\mathcal{A}$ if and only if E_ℓ holds and X_p meets $\mathcal{A}_\ell$.

78. For a natural number n and a rational q, consider the set

$$\{\alpha : |x + \alpha y - q| < 1/n \text{ for some } (x, y) \in P\}.$$

79. Call a sub-box $C = C_1 \times \cdots \times C_n$ *odd* if it has an odd number of elements, i.e. if $|C_i|$ is odd for every i. Given a partition of A into non-trivial boxes B, associate with each B the set O_B of odd sub-boxes of A meeting B in an odd sub-box. Make use of these collections O_B to solve the problem.

81. Suppose that the complete graph on $[n] = \{1, 2, \ldots, n\}$ is decomposed into complete bipartite graphs $G_1, G_2, \ldots, G_r$. Assign a variable X_i to each vertex i, and express this decomposition as a sum of r summands giving the homogeneous polynomial of degree $X_1 X_2 + X_1 X_3 + X_1 X_4 + \cdots + X_{n-1} X_n$.

82. Consider the rank of the matrices.

83. What is the number of maps and what is the number of patterns?

84. Use complex numbers.

86. The main task is to prove the first inequality. To do so, note that there is an unavoidable set X of cardinality $n_k(m)$ that consists of words of length m, and no word of length $2m$ of the form $w^2 = ww$ avoids X.

87. To prove that $d(n) = n^{o(1)}$, construct the prime factorization of n by starting with 1 and then adding to the current number the prime factors one by one. At each step, follow the change in $\log n$ and $\log d(n)$. Thus, at the first step, when 1 is replaced by p, we increase 0 to $\log p$, and 0 to $\log 2$. When we include a prime p which appears to power $a - 1$ in the current value of n, then $\log n$ is again increased by $\log p$, but $\log d(n)$ is increased only by $\log((a + 1)/a)$. Note that, given any $\varepsilon > 0$, there are only finitely many pairs (p, a) for which $\log((a + 1)/a)$ is at least ε times $\log p$.

88. Suppose that the assertion is false, and count the number of paths in two different ways.

89. Use A to define a graph of order n, and apply the result in the previous problem.

90. Note that for constants ξ_i we have

$$\left\| f - \sum_{i=1}^{n} \xi_i \varphi_i \right\|^2 = \left(f - \sum_{i=1}^{n} \xi_i \varphi_i, f - \sum_{i=1}^{n} \xi_i \varphi_i \right) \geq 0.$$

Expand the inner product above and choose suitable values for the constants ξ_i.

93. The answer is that $A(n) = 2^{n-1}$ and $B(n) = 2^n - 1$. To prove this, check and make use of the following two facts:

(i) the process is reversible: if a configuration X can be changed into Y then Y can be changed into X;
(ii) if $X \in V(n)$ and $Y \in V(n - |X|)$ then $X \cup (x + Y) \in V(n)$ for every $x \in X$, where, as usual, $x + Y = \{x + y : y \in Y\}$.

97. Let X_p and Y_p be independent p-random subsets of $[n]$. Note that $X_p \cap Y_p$ is a p^2-random subset of $[n]$.

100. Let us use the numbers $1, \ldots, 6$ to denote the students. With the numbers on the foreheads, the teacher permutes the students: she puts them in some order. For example, the order (i.e. permutation) 512643 means that student #5 gets the highest number, #1 the second highest, etc. What a student can see is a part of this permutation, the part without himself. For example, the student #2 can see the permutation 51643. The students have to come up with the partition $\{2, 4, 5\} \cup \{1, 3, 6\}$. The solution screams out for parity, therefore parity of permutations. This is the signature of a permutation.

The *signature* or *sign* $\text{sgn}(\sigma)$ of a permutation σ of $[n]$ is -1 raised to the number of inversions of σ, i.e. the number of pairs (i, j) such that $1 \leq i < j \leq n$ and $\sigma(i) > \sigma(j)$. A permutation is *even* if its sign is $+1$, and *odd* if its sign is

−1. The set of all even permutations of $[n]$ is the *alternating group A_n* with $n!/2$ elements, i.e. half of the elements of the *symmetric group S_n* of all permutations. The hope has to be that the students can define permutations whose signs give the appropriate partition.

106.   Use (ii) to prove (iii).

109.   Express the ratios in terms of the areas of triangles ABO, BCO and CAO.

110.   Suppose first that an optimal set of n weights can be obtained by adding a weight to an optimal set of $n − 1$ weights. Then what are the weights?

111.   Prove that, for a prime p and integer $\alpha \geq 1$, there are $2^{\alpha-1}$ ways of partitioning a weight of $n = p^\alpha − 1$ lb. so as to be able to weigh, in only one manner, any weight of an integral number of pounds from 1 to n inclusive, it being only permissible to place the weights in one scale-pan. In particular, for $p = 2$ and $n = 5$, there are $2^4 = 16$ ways. Show that every suitable partition corresponds to a polynomial identity.

112.   Call two sequences of reals *equivalent* if they agree in all but finitely many terms. This is clearly an equivalence relation, so every sequence determines an equivalence class. The players agree in a *canonical representative* of each equivalence class, i.e. given an equivalence class E, they pick a canonical sequence z_E in it. After this strong hint the problem is perhaps too easy.

113.   This time we need more equivalence classes, canonical representatives and thresholds than in Problem 112.

114.

118.   To prove (i), apply the Sine Rule to the triangles OAC, OAD, OBC and OBD, and appropriate angles.

119. Denote by J the intersection of the sides CD and EF, and by K the intersection of the sides DE and FA. Make use of the basic properties of cross-ratios to prove that $I[EJFH] = I[EJFG]$.

120. Let $A_0 \cdots A_5$ be our hexagon, and let $A_0 A_1$ meet $A_3 A_4$ in P_0, let $A_1 A_2$ meet $A_4 A_5$ in P_1, and let $A_2 A_3$ meet $A_5 A_0$ in P_2. Finally, let the circle through A_1, P_1 and A_4 meet the line through A_0, A_1 and P_0 in B_0, and the line through A_4, A_3 and P_0 in B_1. Show that the triangles $A_0 A_3 P_2$ and $B_0 B_1 P_1$ are perspective, i.e. their corresponding sides are parallel.

125. Use cross-ratios: see Problem 118.

126. To prove the second recurrence, apply Euler's Pentagonal Theorem from Problem 74. [Seeing the form of this recurrence, this is not much of a hint.]

127. Apply induction on n: in your argument make use of the recurrence relation

$$np(n) = \sum_{v=1} \sum_{k=1} vp(n - kv)$$

from the previous problem, and the inequality that

$$e^{-x}/(1 - e^{-x})^2 < 1/x^2.$$

If you do use this inequality, prove it first.

128. Make use of the fact that if $Z = \{z_1, z_2, \ldots\}$ a countable dense set of vectors in ℓ^1 and $Z' = \{z'_1, z'_2, \ldots\}$ is such that $d(z_i, z'_i) = ||z_i - z'_i|| \to 0$ as $i \to \infty$, then Z' is also dense in ℓ^1.

The Solutions

1. Real Sequences – An Interview Question

(i) *Let $n \geq 1$ be a fixed natural number and let $0 < x_1 < \cdots < x_N < 2n + 1$ be real numbers such that $|kx_i - x_j| \geq 1$ for all natural numbers i, j and k with $1 \leq i < j \leq N$. Then $N \leq n$.*

(ii) *Let $n \geq 1$ be a fixed natural number and let $0 < x_1 < \cdots < x_N < (3n+1)/2$ be real numbers such that $|kx_i - x_j| \geq 1$ for all natural numbers i, j and k with $1 \leq i < j \leq N$ and $k \geq 1$ odd. Then $N \leq n$.*

Proof. (i) Set $x = x_N$. For every i, $1 \leq i \leq N$, let $k_i \geq 0$ be the unique integer such that $x/2 < 2^{k_i} x_i \leq x$. Then $2^{k_i} x_i \geq x/2 + 1/2$ since otherwise $|2^{k_i+1} x_i - x_N| < 1$. Also, $|2^{k_i} x_i - 2^{k_j} x_j| \geq 1$ for all i, j with $1 \leq i < j \leq N$, since $|2^{k_i} x_i - 2^{k_j} x_j| < 1$ would imply

$$|2^{k_i - k_j} x_i - x_j| = 2^{-k_j} |2^{k_i} x_i - 2^{k_j} x_j| < 2^{-k_j} \leq 1.$$

Hence

$$x/2 + 1/2 \leq 2^{k_i} x_i \leq x$$

for every i and $|2^{k_i} x_i - 2^{k_j} x_j| \leq 1$ if $i \neq j$. Consequently,

$$x - (x/2 + 1/2) \geq N - 1, \quad \text{i.e.} \quad 2n + 1 > x_N = x \geq 2N - 1,$$

so $N \leq n$, as claimed.

(ii) We shall copy the proof of the first part verbatim: the only change is that we replace 2 by 3. Thus, set $x = x_N$ so that $2x < 3n + 1$, and for every i, $1 \leq i \leq N$, let $k_i \geq 0$ be the unique integer such that $x/3 < 3^{k_i} x_i \leq x$. Then $3^{k_i} x_i \geq x/3 + 1/3$ as otherwise $|3^{k_i+1} x_i - x_N| < 1$. Also, $|3^{k_i} x_i - 3^{k_j} x_j| \geq 1$ for all i, j with $1 \leq i < j \leq N$, since otherwise

$$|3^{k_i - k_j} x_i - x_j| = 3^{-k_j} |3^{k_i} x_i - 3^{k_j} x_j| < 3^{-k_j} \leq 1.$$

Hence

$$x/3 + 1/3 \le 3^{k_i} x_i \le x$$

for every i and $|3^{k_i} x_i - 3^{k_j} x_j| \le 1$ if $i \ne j$. Consequently,

$$x - (x/3 + 1/3) \ge N - 1, \quad \text{i.e.} \quad 3n + 1 > 2x_N = 2x \ge 3N - 2,$$

so $N \le n$, as claimed. □

Notes. The results above are sharp. For example, if in (i) we weaken the strict inequality $x_N < 2n + 1$ to $x_N \le 2n + 1$ then N can be as large as $n + 1$. Indeed, the $n + 1$ integers $n + 1 < n + 2 < \cdots < 2n + 1$ are such that none is at distance less than 1 from a multiple of another.

The alert reader must have realized that part (ii) holds in greater generality. We postulated that the multiplier k was *odd*, but what we used was that it was at least 3. Clearly, the proof above (given twice, with tiny changes) applies to whatever we take instead of the bounds 2 and 3 above.

This problem is an extension of a basic 'Erdős Problem for Epsilons', namely Problem 2(i) in CTM, a problem Erdős invented and asked from clever students in their early teens. It would have been an ideal question when interviewing candidates for admission to Trinity College, but I had stopped interviewing years before I thought of this problem.

2. Vulgar Fractions – Sylvester's Theorem

Every rational number strictly between 0 and 1 is the sum of a finite number of reciprocals of distinct natural numbers.

Proof. In Sylvester's termonology, the statement is that every vulgar rational number is the sum of a finite number of distinct simple rational numbers. Here a rational number r is *simple* if $r = 1/n$ for some natural number $n \geq 2$, and is *vulgar* if it is in its lowest form $r = a/b$, where $2 \leq a < b$.

We shall prove more, namely that the greedy algorithm produces the required representation: starting with $r = a/b$, where $1 \leq a < b$, let $n_1 \geq 2$ be the minimal natural number satisfying $r \geq 1/n_1$; if this is an equality, we stop the sequence, otherwise we write n_2 for the minimal natural number satisfying $r \geq 1/n_1 + 1/n_2$; if this is an equality, we stop the sequence, otherwise we write n_3 for the minimal natural number satisfying $r \geq 1/n_1 + 1/n_2 + 1/n_3$, etc. In this construction we must have

$$\frac{1}{n_i - 1} > \frac{1}{n_i} + \frac{1}{n_{i+1}},$$

since otherwise we could and so would have taken $n_i - 1$ instead of n_i. Hence

$$n_{i+1} \geq n_i(n_i - 1) + 1;$$

in particular, the sequence (n_i) is increasing: $2 \leq n_1 < n_2 < \cdots$. Therefore, all we have to check is that this process stops after a finite number of steps.

For the fraction $\frac{4699}{7320}$ this goes as follows:

$$\frac{4699}{7320} = \frac{1}{2} + \frac{1039}{7320} = \frac{1}{2} + \frac{1}{8} + \frac{124}{7320} = \frac{1}{2} + \frac{1}{8} + \frac{31}{1830} = \frac{1}{2} + \frac{1}{8} + \frac{1}{60} + \frac{1}{3660}.$$

The inspiration we should get from this expansion is that in the sequence of the remainders,

$$\frac{4699}{7320} > \frac{1039}{7320} > \frac{31}{1830} > \frac{1}{3660},$$

49

the numerators are strictly decreasing. If this is the case in general then this is the only property we have to check to conclude that the sequence $n_1 < n_2 < \cdots$ is finite and so gives us an expansion of $r = a/b$ into a sum of simple fractions.

To check this, let us look again at the sequence $2 \leq n_1 < n_2 < \cdots$. Starting with $r = a/b$, set $a_0 = a$ and $b_0 = b$, so that $1 \leq a_0 < b_0$. Assume that we have found a_i/b_i and $2 \leq n_1 < n_2 < \cdots < n_i$ such that

$$\frac{a_i}{b_i} = \frac{a}{b} - \frac{1}{n_1} - \cdots - \frac{1}{n_i}.$$

If $a_i = 0$, stop the process and note that

$$\frac{a}{b} = \frac{1}{n_1} + \cdots + \frac{1}{n_i}.$$

Otherwise let n_{i+1} be the minimal natural number such that

$$a_i n_{i+1} \geq b_i,$$

i.e. let $n_{i+1} = \lceil b_i/a_i \rceil$ be the unique natural number satisfying

$$0 \leq a_i n_{i+1} - b_i < a_i,$$

and set

$$a_{i+1} = a_i n_{i+1} - b_i < a_i \text{ and } b_{i+1} = b_i n_{i+1}.$$

Then

$$\frac{a_{i+1}}{b_{i+1}} = \frac{a}{b} - \frac{1}{n_1} - \cdots - \frac{1}{n_{i+1}}.$$

Since $a_0 > a_1 > \cdots > a_{i+1}$, our process does stop after finitely many steps and we have our representation.

As another illustration of our greedy algorithm, take $r = 335/336$. Then $n_1 = \lceil 336/335 \rceil = 2$, $a_1 = 334$, $b_1 = 672$, $n_2 = \lceil 672/334 \rceil = 3$, $a_2 = 330$, $b_2 = 2016$, $n_3 = \lceil 2016/330 \rceil = 7$, $a_3 = 294$, $b_3 = 14112$, $n_4 = \lceil 14112/294 \rceil = 14112/294 = 48$, so

$$\frac{335}{336} = \frac{1}{2} + \frac{1}{3} + \frac{1}{7} + \frac{1}{48}. \qquad \square$$

Notes. This little result is due to J.J. Sylvester (1814–1897), the great algebraist of the 19th century, from whom we borrowed the title of the exercise as well. [I hope that it is clear to every reader that this is *not* one of the many results that made Sylvester famous.]

Sylvester was a Fellow of St John's College, Cambridge, and was one of the first major mathematicians who worked (for some years) in the USA: appropriately, as the inaugural professor of mathematics at the newly founded Johns Hopkins University. To see how times have changed, it is interesting to recall that Sylvester did not have much faith in the dollar, so he demanded to be paid in gold. One of his lasting legacies in the States is the *American Journal of Mathematics* he founded in 1878. (Actually, he published this paper in that journal.) A few years later he returned to England to take up a professorship at Oxford.

The expansion above of a rational r, $0 < r < 1$, as a sum of reciprocals of natural numbers is usually called the *Sylvester representation* of r. A totally pedestrian (but more informative) name for it is the name we have used, *greedy representation*. Let us describe it again, emphasizing its 'greedy' aspect. We choose $n_1 < n_2 < \cdots$ one after the other. Having defined n_i, if

$$r - \frac{1}{n_1} - \cdots - \frac{1}{n_{i-1}} - \frac{1}{n_i} > 0,$$

we choose $1/n_{i+1}$ as large as possible conditional on

$$r - \frac{1}{n_1} - \cdots - \frac{1}{n_i} - \frac{1}{n_{i+1}} \geq 0.$$

Since our conditions did not allow us to use $1/(n_i - 1)$ instead of $1/n_i$, we have

$$r - \frac{1}{n_1} - \cdots - \frac{1}{n_{i-1}} - \frac{1}{n_i - 1} < 0,$$

and so

$$\frac{1}{n_i - 1} - \frac{1}{n_i} > \frac{1}{n_{i+1}},$$

i.e.

$$n_{i+1} \geq n_i(n_i - 1) + 1.$$

Occasionally, this lower bound on the growth of the sequence (n_i) turns out to be handy.

References

Sylvester, J.J., On a point in the theory of vulgar fractions, *Amer. J. Math.* **3** (1880) 332–335.

Sylvester, J.J., Postscript to a note on a point in vulgar fractions, *Amer. J. Math.* **3** (1880) 388–389.

3. Rational and Irrational Sums

Let $2 \leq n_1 < n_2 < \cdots$ be a sequence of positive integers such that

$$n_{i+1} \geq n_i(n_i - 1) + 1 \tag{1}$$

for every $i \geq 1$. Then

$$r = \sum_{i=1}^{\infty} \frac{1}{n_i}$$

is rational if and only if in (1) *equality holds for all but finitely many values of i.*

Proof. Let us make some preliminary observations. Let $2 \leq a_h < a_{h+1} < \cdots$ be such that

$$a_{i+1} = a_i(a_i - 1) + 1,$$

i.e.

$$\frac{1}{a_i} = \frac{1}{a_i - 1} - \frac{1}{a_{i+1} - 1}$$

for every $i \geq h$. Then, for $k \geq h$, we have

$$\sum_{i=h}^{k} \frac{1}{a_i} = \sum_{i=h}^{k} \left(\frac{1}{a_i - 1} - \frac{1}{a_{i+1} - 1} \right) = \frac{1}{a_h - 1} - \frac{1}{a_{k+1} - 1},$$

and so

$$\sum_{i=h}^{\infty} \frac{1}{a_i} = \frac{1}{a_h - 1}. \tag{2}$$

Relation (2) tells us that r is well defined: the series $\sum_{i=1}^{\infty} 1/n_i$ is indeed convergent. Also, $r = 1$ if $n_1 = 2$ and $n_{i+1} = n_i(n_i - 1) + 1$ for every i, otherwise $0 < r < 1$.

53

Let us turn to the actual assertions of the exercise.

(i) First, suppose that in (1) equality holds for all but finitely many values of i. This means that there is an integer k such that in (1) equality holds for $i \geq k$. Then, by (2),

$$r = \sum_{i=1}^{k-1} \frac{1}{n_i} + \sum_{i=k}^{\infty} \frac{1}{n_i} = \sum_{i=1}^{k-1} \frac{1}{n_i} + \frac{1}{n_k - 1},$$

so r is indeed a rational number.

(ii) Second, suppose that in (1) strict inequality holds for infinitely many values of i. Then, by (2),

$$\sum_{i=h}^{\infty} \frac{1}{n_i} < \frac{1}{n_h - 1}$$

for every $h \geq 1$. This tells us that

$$\frac{1}{n_h} < r - \sum_{i=1}^{h-1} \frac{1}{n_i} = \sum_{i=1}^{\infty} \frac{1}{n_i} - \sum_{i=1}^{h-1} \frac{1}{n_i} = \sum_{i=h}^{\infty} \frac{1}{n_i} < \frac{1}{n_h - 1}.$$

Consequently, recalling Problem 2, the original sequence (n_i) gives the Sylvester representation of r:

$$r = \sum_{i=1}^{\infty} \frac{1}{n_i}.$$

Since for a rational number r the Sylvester representation if finite, r is indeed irrational, completing the proof. $\square$

Note. This simple exercise is from a 1963 paper of Erdős and Stein.

Reference

Erdős, P. and S. Stein, Sums of distinct unit fractions, *Proc. Amer. Math. Soc.* **14** (1963) 126–131.

4. Ships in Fog

Five ships, A, B, C, D and E, are sailing in a fog with constant and different speeds, and constant and different straight-line courses, with different directions. The seven pairs AB, AC, AD, BC, BD, CE and DE have each had near misses, call them 'collisions'. Then, in addition, E collides with A and B, and C collides with D.

Proof. Represent the system in the 3-dimensional space $\mathbb{R}^3$ by plotting the graphs of position of a ship against time: let the position of ship A at time t be $(x_a(t), y_a(t))$, so that $\{(x_a(t), y_a(t), t) : -\infty < t < \infty\}$ is the 'world line' a of A. Define the world lines b, c, d and e analogously.

Two ships collide if their world lines meet. Now, the world lines a, b and c meet, so these lines are in the same plane P. If at least two points of a line are in P, then the entire line is in P. Hence the world line d is in P, as D collides with A and B. Furthermore, e is in P since E collides with C and D. Consequently all five lines a, b, c, d and e are in P, and as no two lines are parallel, every pair of lines meet. $\square$

Note. This is a slight variant of a simple problem in the 'Mathematics with Minimum Raw Material' chapter of *Littlewood's Miscellany*.

Reference

Littlewood, J.E., *Littlewood's Miscellany*. Edited and with a foreword by Béla Bollobás, Cambridge University Press (1980).

5. A Family of Intersections

For $0 < q < 1$, the q-probability of a family $\mathcal{F}$ of subsets of $[n] = \{1, \ldots, n\}$ is

$$\mathbb{P}_q(\mathcal{F}) = \sum_{F \in \mathcal{F}} \mathbb{P}_q(F) = \sum_{F \in \mathcal{F}} q^{|F|} (1 - q)^{n-|F|}.$$

Let $\mathcal{A}$ be a family of subsets of $[n]$ with p-probability r, i.e. let $\mathbb{P}_p(\mathcal{A}) = r$, and define

$$\mathcal{J} = \mathcal{J}(\mathcal{A}) = \{A \cap B : \ A, B \in \mathcal{A}\}.$$

Then

$$\mathbb{P}_{p^2}(\mathcal{J}) \geq r^2.$$

Proof. Let X_p and Y_p be independent p-random subsets of $[n]$. Then $X_p \cap Y_p$ is a p^2-random subset of $[n]$ since the events $\{1 \in X_p \cap Y_p\}, \{2 \in X_p \cap Y_p\}, \ldots, \{n \in X_p \cap Y_p\}$ are independent, and each has probability p^2. Consequently, if $\mathcal{B}$ is a family of subsets of $[n]$, then its p^2-probability is

$$\mathcal{P}_{p^2} = \mathcal{P}(X_p \cap Y_p \in \mathcal{B}).$$

Applying this identity to $\mathcal{B} = \mathcal{I}$, we find that

$$\mathbb{P}_{p^2}(\mathcal{J}) = \mathbb{P}(X_p \cap Y_p \in \mathcal{J}) \geq \mathbb{P}(X_p \in \mathcal{A} \text{ and } Y_p \in \mathcal{A})$$
$$= \mathbb{P}(X_p \in \mathcal{A}) \, \mathbb{P}(Y_p \in \mathcal{A}) = r^2,$$

as claimed. $\qquad\qquad\qquad\qquad\qquad\qquad\qquad\qquad\qquad\qquad\qquad\qquad\qquad\square$

Notes. This inequality was used by Ellis and Narayanan in their beautiful paper in which they proved the old conjecture of Peter Frankl that if $\mathcal{A} \subset \mathcal{P}_n$ is a *symmetric* 3-wise intersecting family of subsets of $[n]$ then $|\mathcal{A}| = o(2^n)$. (Thus $A \cap B \cap C \neq \emptyset$ for all $A, B, C \in \mathcal{A}$ and the automorphism group of $\mathcal{A}$ is transitive on $[n]$, i.e. for all $1 \leq i < j \leq n$ there is a permutation of $[n]$ that maps i into j, and maps every set in $\mathcal{A}$ into a set in $\mathcal{A}$.) More precisely, Ellis and Narayanan proved this for $p = 1/2$: sitting in a seminar Narayanan gave on the proof of Frankl's conjecture, Paul Balister noticed this extension and the lovely proof given above.

The proof trivially carries over to the following extension. Let $\mathcal{A}_1, \ldots, \mathcal{A}_k$ be families of subsets of $[n]$, and $0 < p_1, \ldots, p_k$, and set

$$\mathcal{J} = \{A_1 \cap \cdots \cap A_k : A_i \in \mathcal{A}\}.$$

Reference

Ellis, D. and B. Narayanan, On symmetric 3-wise intersecting families, *Proc. Amer. Math. Soc.* **145** (2017) 2843–2847.

6. The Basel Problem – Euler's Solution

The sum of the infinite series

$$\sum_{k=1}^{\infty} 1/k^2 = 1 + \frac{1}{4} + \frac{1}{9} + \frac{1}{16} + \cdots$$

is $\pi^2/6$.

Proof. The infinite power series

$$1 - x^2/3! + x^4/5! - x^6/7! \pm \cdots$$

is convergent for every $x \in \mathbb{R}$; we shall view the function $p(x)$ it gives,

$$p(x) = 1 - x^2/3! + x^4/5! - x^6/7! \pm \cdots ,$$

as a *polynomial.* Since

$$e^{ix} = \cos x + i \sin x = 1 + ix - x^2/2! - ix^3/3! + x^4/4! + ix^5/5! \pm \cdots ,$$

we have

$$\sin x = (e^{ix} - e^{-ix})/(2i) = x - x^3/3! + x^5/5! - x^7/7! \pm \cdots ,$$

so

$$p(x) = \frac{x - x^3/3! + x^5/5! - x^7/7! \pm \cdots}{x} = \frac{\sin x}{x}.$$

Consequently, the roots of $p(x)$ are the roots of $\sin x$, except for 0, i.e. $\pm\pi, \pm 2\pi, \pm 3\pi, \cdots$. Also, every root has multiplicity one, because $\sin x$ has no double roots. Now, using the roots of the 'polynomial' $p(x)$ to factorize it, we have

$$p(x) = \left(1 - \frac{x}{\pi}\right)\left(1 + \frac{x}{\pi}\right)\left(1 - \frac{x}{2\pi}\right)\left(1 + \frac{x}{2\pi}\right)\cdots$$

$$= \left(1 - \frac{x^2}{\pi^2}\right)\left(1 - \frac{x^2}{4\pi^2}\right)\cdots .$$

And now let us expand this infinite product to get another power series representation of $p(x)$:

$$p(x) = 1 - \frac{x^2}{\pi^2}\left(1 + \frac{1}{4} + \frac{1}{9} + \frac{1}{16} + \frac{1}{25} + \cdots\right) + \cdots .$$

where we do not care about the coefficients of the higher powers of x. This power series must agree term by term with our earlier power series for $p(x)$; in particular, the coefficients of x^2 are the same:

$$-\frac{1}{6} = -\frac{1}{\pi^2}\left(1 + \frac{1}{4} + \frac{1}{9} + \frac{1}{16} + \frac{1}{25} + \cdots\right),$$

whose rearrangement is the required identity. □

Notes. The Basel Problem of finding the numerical value of the infinite series $1 + \frac{1}{2^2} + \frac{1}{3^2} + \cdots$ will forever be associated with Leonhard Euler. In fact, it is likely that this problem was posed by the Bologna clergyman and mathematician Pietro Mengoli around 1650, and became famous as the *Basel Problem*, when Jacob Bernoulli, who was born in Basel and became a professor of mathematics there, attempted to solve it and failed. What he could show was that the sum is less than 2. Just about all the great mathematicians in Europe (including Leibniz, de Moivre, Johann and Daniel Bernoulli, and Goldbach) tried to solve it, but at best could only suggest approximations. Euler, who was also born in Basel and studied under Johann Bernoulli, became fascinated by this problem in his early twenties, and eventually solved it in 1735, when he was twenty-eight: this solution gave him instant fame. Later he gave at least two more proofs that are easy to make entirely rigorous.

The interest today in this problem is startling: after all, this is a really old chestnut. Still, it seems that mathematicians and amateurs alike remain fascinated by this beautiful problem, which does lend itself to an astonishing variety of proofs. A bit like the theorem of Pythagoras, but the sophistication of the Basel Problem is incomparably greater. In 1999, Robin Chapman of Exeter wrote a manuscript entitled 'Evaluating $\zeta(2)$', in which he collected fourteen proofs of this identity. Although in 2003 he updated this MS, sadly, he never submitted it for publication. Since this MS was written, several other proofs have been published. For a meagre selection of proofs, see the references below.

References

Apostol, T.M., A proof that Euler missed: Evaluating $\zeta(2)$ the easy way, *Math. Intelligencer* **5** (1983) 59–60.

Chapman, R., Evaluating $\zeta(2)$, unpublished manuscript, 13 pp (2003).

Dunham, W., *Euler: The Master of Us All*, The Dolciani Mathematical Expositions **22**, MAA (1999).

Dunham, W., When Euler met l'Hôpital, *Math. Mag.* **82** (2009) 16–25.

Giesy, D.P., Still another elementary proof that $\sum_{k=1}^{\infty} 1/k^2 = \pi^2/6$, *Math. Mag.* **45** (1972) 148–149.

Harper, J.D., Another simple proof of $1 + \frac{1}{2^2} + \frac{1}{3^2} + \cdots = \pi^2/6$, *Amer. Math. Monthly* **110** (2003) 540–541.

Hofbauer, J., A simple proof of $1 + \frac{1}{2^2} + \frac{1}{3^2} + \cdots = \pi^2/6$ and related identities, *Amer. Math. Monthly* **109** (2002) 196–200.

Kalman, D., Six ways to sum a series, *College Math. J.* **24** (1993) 402–421.

Pace, L., Probabilistically proving that $\zeta(2) = \pi^2/6$, *Amer. Math. Monthly* **118** (2011) 641–643.

Papadimitriou, I., A simple proof of the formula $\sum_{k=1}^{\infty} k^{-2} = \pi^2/6$ *Amer. Math. Monthly* **80** (1973) 424–425.

7. Reciprocals of Primes – Euler and Erdős

The sum of the reciprocals of the primes is divergent.

Proof. We shall give three simple, elegant and famous proofs. As customary, in our formulae p will denote a prime; thus, $\sum_{p \leq n}$ denotes summation over all primes that are at most n. Also, we write $p_1 = 2 < p_2 = 3 < \ldots$ for the sequence of primes.

Euler's Proof Since every natural number is a product of primes,

$$\sum_{k=1}^{n} 1/k \leq \prod_{p \leq n} \left(1 + 1/p + 1/p^2 + \cdots \right) = \prod_{p \leq n} (1 - 1/p)^{-1}.$$

Furthermore, as $\log n \leq \sum_{k=1}^{n} 1/k$, we have

$$\prod_{p \leq n} (1 - 1/p)^{-1} \geq \log n.$$

Very crudely, if $0 \leq t \leq 1/2$ then $\log(1 - t) \geq -2t$, so taking the logarithm of this inequality, we find that

$$2 \sum_{p \leq n} 1/p \geq \log \log n,$$

completing Euler's proof. □

Erdős's First Proof Suppose that $\sum_1^{\infty} 1/p_i < \infty$, and let k be such that $\sum_{i=k+1}^{\infty} 1/p_i < 1/2$. Given a natural number n, let A_n be the set of integers at most n not divisible by any of the 'large' primes $p_{k+1}, p_{k+2}, \ldots$, and let B_n be the rest of the integers up to n. Then every integer $m \in A_n$ is of the form $m = st^2$, where s is square-free, i.e. the product of some of the 'small' p_i, each with exponent 1. Since every such square-free number s corresponds to a

61

subset of the set $\{p_1, \ldots, p_k\}$, we have at most 2^k choices for s. Furthermore, as $t^2 \le n$, we have at most $\sqrt{n}$ choices for t, so

$$|A_n| \le 2^k \sqrt{n}.$$

Since every integer $m \ge 1$ has $\lfloor n/m \rfloor$ multiples up to n, and every integer $m \in B_n$ is the multiple of at least one of the 'large' primes $p_{k+1}, p_{k+2}, \ldots,$

$$|B_n| \le \sum_{i=k+1}^{\infty} \lfloor n/p_i \rfloor \le \sum_{i=k+1}^{\infty} n/p_i = n \sum_{i=k+1}^{\infty} 1/p_i < n/2.$$

Putting together our bounds for $|A_n|$ and $|B_n|$ we find that

$$n = |A_n| + |B_n| < 2^k \sqrt{n} + n/2,$$

which is a contradiction if $n \ge 2^{2(k+1)}$. □

Erdős's Second Proof As in the first proof, assume that $\sum_1^\infty 1/p_i < \infty$, but this time let k be such that the tail of this series is even smaller:

$$\sum_{i=k+1}^{\infty} \frac{1}{p_i} < \frac{1}{8}.$$

Also, let us note a simple inequality following from a telescoping series:

$$\sum_{m=2}^{\infty} \frac{1}{m^2} \le \frac{1}{4} + \frac{1}{2 \cdot 3} + \frac{1}{3 \cdot 4} + \cdots = \frac{1}{4} + \left(\frac{1}{2} - \frac{1}{3}\right) + \left(\frac{1}{3} - \frac{1}{4}\right) + \cdots = \frac{3}{4}.$$

If n is large enough, these two inequalities give us a contradiction. Indeed, every number up to n is either a multiple of a square $m^2 \ge 4$, or is a multiple of a 'large' prime, a prime p_i with $i > k$, or is a product of some of the 'small' primes, each with exponent 1. Hence,

$$n \le n/8 + 3n/4 + 2^k,$$

which is a contradiction if $n > 2^{k+3}$. □

Notes. This is a very well-known result, one of the basic results about the distribution of primes. The result is due to Euler, who gave a not entirely complete proof of it in 1737. In 1938 Erdős published the two proofs above; a few years later the floodgates opened and numerous other proofs appeared in print: see Bellman (1943), Dux (1956), Moser (1958), Clarkson (1966) and Meštrović (2013) for a selection of these.

In the first proof above, the one we attributed to Euler, we were far too generous with our bounds: with a little more care we could have got

$$\sum_{p \leq n} \frac{1}{p} \geq (1 + o(1)) \log \log n.$$

Euler's proof is connected to *Euler's product formula* for the *Riemann zeta function*:

$$\zeta(s) = \sum_{n=1}^{\infty} \frac{1}{n^s} = \prod_p \frac{1}{1 - p^{-s}}$$

for all $s \in \mathbb{C}$ with $\mathrm{Re}\, s > 1$.

In their book on discrete harmonic analysis, Ceccherini-Silberstein, Scarabotti and Tolli presented Euler's proof as follows (see pp. 97–98). For $s > 1$,

$$\log \zeta(s) = \sum_p \log \left(\frac{1}{1 - p^{-s}} \right) = \sum_p \left(\frac{1}{p^s} + R(\frac{1}{p^s}) \right),$$

where $|R(1/p^s)| < 1/p^{2s}$. This bound on R implies that

$$\left| \sum_p R \left(\frac{1}{p^s} \right) \right| \leq \sum_p \frac{1}{p^{2s}} \leq \sum_{n=1}^{\infty} \frac{1}{n^2} = \frac{\pi^2}{6}.$$

Therefore,

$$\sum_p \frac{1}{p^s} \geq \log \zeta(s) - \frac{\pi^2}{6},$$

which tends to $+\infty$ for $s \to 1^+$, since $\zeta(s) = \sum_{n=1}^{\infty} 1/n^s$ tends to $+\infty$ for $s \to 1^+$.

Concerning the two proofs Erdős gave, it is interesting to note that he published both in the same paper of less than a page and a half: although by then he had been in Manchester for several years, this brief note is in German.

References

Bellman, R., A note on the divergence of a series, *Amer. Math. Monthly* **50** (1943) 318–319.

Ceccherini-Silberstein, T., F. Scarabotti and F. Tolli, *Discrete Harmonic Analysis – Representations, Number theory, Expanders, and the Fourier Transform*, Cambridge Studies in Advanced Mathematics **172**, Cambridge University Press (2018).

Clarkson, J.A., On the series of prime reciprocals, *Proc. Amer. Math. Soc.* **17** (1966) 541.

Dux, E., Ein kurzer Beweis der Divergenz der unendlichen Reihe $\sum_{r=1}^{\infty} 1/p_r$ (in German), *Elem. Math.* **11** (1956) 50–51.

Erdős, P., Über die Reihe $\sum \frac{1}{p}$ (in German), *Mathematica, Zutphen B* **7** (1938) 1–2.

Euler, L., Variae observationes circa series infinitas, *Comment. Acad. Sci. Petropol.* **9** (1744) 160–188. [Reprinted in *Opera Omnia* I.14, 216–244, Teubner, Lipsiae et Berolini, 1924.]

Meštrovič, R., A note on two Erdős's proofs of the infinitude of primes, *Electronic Notes in Discrete Mathematics* **43** (2013) 179–186.

Moser, L., On the series $\sum 1/p$, *Amer. Math. Monthly* **65** (1958) 104–105.

8. Reciprocals of Integers

Let $1 < n_1 < n_2 < \cdots$ be a sequence of natural numbers such that $\sum_{i=1}^{\infty} 1/n_i < \infty$. Then the set

$$M = M(n_1, n_2, \ldots) = \{n_1^{\alpha_1} \cdots n_k^{\alpha_k} : \alpha_i \geq 0\}$$

has zero density.

Proof. Write M_n for the set of integers in M not larger than n. Let $\varepsilon > 0$, so that our task is to show that $|M_n| < \varepsilon n$ if n is large enough. To this end, let k be such that $\sum_{i=k+1}^{\infty} 1/n_i < \varepsilon/2$. Then the number of integers up to n that are divisible by at least one n_i for $i \geq k+1$ is at most

$$\sum_{i=k+1}^{\infty} n/n_i < \varepsilon n/2.$$

Every other integer in M_n is of the form

$$n_1^{\alpha_1} \cdots n_k^{\alpha_k} = t^2 \prod_{i \in I} n_i,$$

where $I \subset [k] = \{1, \ldots, k\}$. There are at most $\sqrt{n}2^k$ such numbers since we have at most $\sqrt{n}$ choices for t and at most 2^k choices for the set I. This tells us that

$$|M_n| < \varepsilon n/2 + \sqrt{n}\, 2^k.$$

Consequently, if $n > 2^{2(k+1)}/\varepsilon^2$ then $|M_n| < \varepsilon n$. □

Notes. Clearly, the assertion in this problem is considerably stronger than the fact we proved in Problem 7, namely that $\sum_p 1/p = \infty$. Also, the proof here is just the first proof of Erdős we presented there, *mutatis mutandis*.

Reference

Erdős, P., Über die Reihe $\sum \frac{1}{p}$, *Mathematica, Zutphen B* **7** (1938) 1–2.

9. Completing Matrices

An $r \times n$ matrix $A_{r,n}$ of zeros and ones has an extension to an $n \times n$ matrix of zeros and ones with precisely k ones in each row and each column if and only if each row has precisely k ones, and in each column there are at least $k + r - n$ and at most k ones.

Proof. The *necessity* of the conditions is rather trivial. In fact, it *is* utterly trivial that each row of $A_{r,n}$ must have precisely k ones and each column must have at most k ones. Only the third condition needs a tiny argument: when extending $A_{r,n}$, each column can gain at most $n - r$ ones, and as we have to end up with k ones, we must have started with at least $k - (n - r) = k + r - n$ ones.

Turning to the proof of the *sufficiency* of the conditions, as suggested in the *Hint*, we shall adjoin rows one by one to our matrix $A_{r,n}$ satisfying the conditions. Let $C_1, \ldots, C_n$ be the columns of $A_{r,n}$, and set

$$A = \{i : C_i \quad \text{has } k + r - n \text{ ones}\},$$
$$B = \{i : C_i \quad \text{has at most } k - 1 \text{ ones}\},$$

so that $A \subset B$. Since kr of the entries of $A_{r,n}$ are ones,

$$kr \leq |A|(k + r - n) + (n - |A|)k$$

and

$$kr \geq (n - |B|)k + |B|(k + r - n).$$

These inequalities imply that

$$|A| \leq k \leq |B|.$$

Consequently, we may add to $A_{r,n}$ a row $(a_1, \ldots, a_n)$ with k ones and $n - k$ zeros such that

$$C = \{i : a_i = 1\}$$

satisfies

$$A \subset C \subset B.$$

Then the obtained $(r + 1) \times n$ matrix $A_{r+1,n}$ of zeros and ones does satisfy the conditions:

(1) each row has precisely k ones,
(2) since $A \subset C$, each column has at least $(k + r - n) + 1 = k + (r + 1) - n$ ones,
(3) since $C \subset B$, each column has at most k ones.

Thus adding such rows one by one, after $n - r$ steps we obtain a matrix with k ones in each row and each column, completing our proof. □

10. Convex Polyhedra – Take One

For every point of a convex polyhedron there is a face such that the orthogonal projection of the point on the plane of the face falls in the interior of the face.

Proof. Suppose there is a point whose orthogonal projection on every face is outside the interior of the face. Then affixing a mass at such a point, and leaving the rest of the polyhedron massless, putting the polyhedron on a plane, we would get a *perpetuum mobile*, perpetual motion machine: a tumbler that keeps working forever. It is obvious that the tumbling continues if the projection is outside the face: if the projection is on an edge then no force is needed to make the polyhedron continue its tumbling motion.

A less romantic way of putting this argument goes as follows. Given a point P of the polyhedron, take a face F at minimal distance from P. Then the projection P' of P on the plane of F is within the interior of F, since otherwise the segment PP' meets another face, and that face is nearer to P than F. This applies to the case when P' is on the edge bounding F from another face G: then G is nearer to P than F. □

Notes. This is just about the simplest mathematical result proved by appealing to physics. I first heard of this problem from Paul Dirac, then years later from

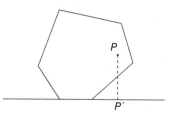
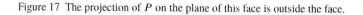

Figure 17 The projection of P on the plane of this face is outside the face.

John Conway, and finally from the master of such arguments, Tadashi Tokieda, two of whose papers we mention below. This is a mathematical joke, so it is rude to say that one knows it.

References

Tokieda, T.F., Mechanical ideas in geometry, *Amer. Math. Monthly* **105** (1998) 697–703.
Tokieda, T.F., Roll models, *Amer. Math. Monthly* **120** (1998) 265–282.

11. Convex Polyhedra – Take Two

Every 3-dimensional polyhedron with at least thirteen faces has a face meeting at least six other faces.

Proof. Let K be a convex polyhedron with V vertices, E edges and F faces. Suppose no face of K meets more than five other faces so, in particular, every face has at most five sides.

If a face is bounded by an ℓ-cycle and the vertices of this ℓ-cycle have degrees $d_1, \ldots, d_\ell$ then this face meets $\sum_{i=1}^{\ell}(d_i - 2)$ other faces, since any two faces are either disjoint or meet in a vertex or meet in an edge. Hence,

$$\sum_{i=1}^{\ell}(d_i - 2) \le 5. \tag{1}$$

Euler's polyhedron theorem tells us that

$$V + F = E + 2.$$

Also, writing f_i for the number of faces with i sides,

$$\sum_i f_i = \sum_{i=3}^{5} f_i = F \qquad \text{and} \qquad \sum_{i=3}^{5} i f_i = 2E,$$

so

$$2E \le 5F.$$

Write $D_1, \ldots, D_V$ for the degrees of the V vertices of K. Summing inequality (1) over all F faces, we find that

$$\sum_{i=1}^{V} D_i^2 - 2\sum_{i=3}^{5} i f_i = \sum_{i=1}^{V} D_i^2 - 4E \le 5F.$$

70

Since $\sum_{i=1}^{V} D_i = 2E$ and the function $x \rightarrow x^2$ is convex, this implies that

$$V(2E/V)^2 - 4E \leq 5F,$$

and so

$$4E^2 \leq (4E + 5F)V = (4E + 5F)(E - F + 2) = 4E^2 - 5F^2 + EF + 8E + 10F.$$

Consequently, since $E \leq 5F/2$,

$$5F^2 \leq EF + 8E + 10F \leq 5F^2/2 + 30F,$$

implying that $F \leq 12$ and so completing our proof. $\square$

Note. Every face of a dodecahedron meets five other faces, so the bound thirteen we have given in our problem is best possible.

12. A Very Old Tripos Problem

Let p, q and r be complex numbers with $pq \neq r$. Transform the cubic $x^3 - px^2 + qx - r = 0$, where the roots are a, b, c, into one whose roots are $\frac{1}{a+b}, \frac{1}{a+c}, \frac{1}{b+c}$.

Solution. First, let us note that $a + b$, $a + c$ and $b + c$ are non-zero. Indeed, if we had $a + b = 0$, say, then we would find that $x^3 - px^2 + qx - r = (x - a)(x + a)(x - c) = x^3 - cx^2 - a^2x + a^2c$, so $pq = r$, contradicting our assumption. Thus $1/(a + b)$, $1/(a + c)$ and $1/(b + c)$ are complex numbers.

Since $x^3 - px^2 + qx - r = (x - a)(x - b)(x - c)$, we have

$$a + b + c = p, \quad ab + bc + ca = q, \quad abc = r.$$

Writing $x^3 - p'x^2 + q'x - r'$ for the transformed cubic with roots $\frac{1}{a+b}, \frac{1}{a+c}, \frac{1}{b+c}$, the coefficients p', q', r' satisfy similar equations. First,

$$p' = \frac{1}{a+b} + \frac{1}{b+c} + \frac{1}{c+a} = \frac{(b+c)(c+a) + (a+b)(c+a) + (a+b)(b+c)}{(a+b)(b+c)(c+a)}.$$

The numerator is

$$(a^2 + 2ab + b^2 + 2ac + 2bc + c^2) + (ab + bc + ca) = (a+b+c)^2 + (ab+bc+ca) = p^2 + q,$$

and the denominator is

$$2abc + a^2(b+c) + b^2(c+a) + c^2(a+b) = (a+b+c)(ab+bc+ca) - abc = pq - r.$$

Similarly,

$$q' = \frac{1}{(a+b)(b+c)} + \frac{1}{(b+c)(c+a)} + \frac{1}{(c+a)(a+b)},$$

so

$$q' = \frac{2p}{pq - r}$$

and

$$r' = \frac{1}{(a+b)(b+c)(c+a)} = \frac{1}{pq-r}.$$

Hence,

$$x^3 - \frac{p^2+q}{pq-r}x^2 + \frac{2p}{pq-r}x - \frac{1}{pq-r}$$

is the cubic we had to find. □

Notes. In 1801 this was the twelfth of fourteen problems on the last paper set to the candidates hoping to become Wranglers. I don't expect that the reader will find the problem exciting, but it is fun to see what counted for a proper Tripos problem over 200 years ago. Not surprisingly, the standard was very low.

13. Angle Bisectors – the Lehmus–Steiner Theorem

If two angle bisectors of a triangle are equal then the angles themselves are also equal.

Proof. Let ABC be a triangle with angles α and β at A and B, and with angle bisectors AD and BE of the same length. Suppose that $\alpha \neq \beta$, say, $\alpha < \beta$. To prove the result, we have to show that this leads to a contradiction.

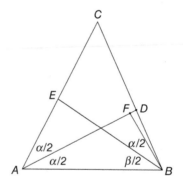

Figure 18 The setup if $\alpha < \beta$.

Let F be the point on AD such that the angle ABF is $(\alpha + \beta)/2 < \pi/2$, and so the angle EBF is $(\alpha + \beta)/2 - \beta/2 = \alpha/2$, as is the angle EAF (see Figure 18). The equality of these two angles implies that the points F, E, A and B are concyclic.

Shorter chords of a circle subtend smaller acute angles; therefore, as AF is shorter than AD, which has the same length as BE, and the angles ABF and EAB are acute, the angle ABF is less than the angle EAB, i.e. $(\alpha + \beta)/2 < \alpha$, contradicting our assumption that $\alpha < \beta$, and so completing our proof. □

Notes. This is a *very* old chestnut, which in my school days was taught to most pupils, but today it is less well known. In my case, it was given to me as an exercise by the geometer István Reiman, to whom I am eternally grateful for teaching me so much mathematics in my teens. As I learned much later, the result was posed as a problem by C.L. Lehmus in 1840, and proved by the great geometer Jacob Steiner.

14. Langley's Adventitious Angles

Let ABC be an isosceles triangle with angle 20° at the apex A. Let D be a point on AB and E a point on AC such that ∠BCD = 50° and ∠CBE = 60°. Then ∠BED = 30°.

Proof. Following the *Hint*, add a point F on the side AC such that $\angle CBF = 20°$. We shall repeatedly make use of the fact that the three angles of a triangle sum to 180°. From the information we have we can find quite a few of the angles determined by our points; in particular, $\angle BCF = \angle BFC = 80°$, $\angle BCD = \angle BDC = 50°$ and $\angle FBE = \angle FEB = 40°$, as in Figure 19. Hence, the corresponding three triangles are isosceles, implying that $BF = BD = FE$, i.e. these three distances are the same.

Since $\angle DBF = 60°$ and $BF = BD$, the triangle BDF is equilateral, and so $DF = FE$, i.e. the triangle FDE is also isosceles. The angle at the apex F of this triangle is

$$\angle DFE = 180° - \angle BFC - \angle BFD = 180° - 80° - 60° = 40°,$$

so the angles at its base are $\angle DEF = \angle EDF = 70°$. Finally,

$$\angle DEB = \angle DEF - \angle BEF = 70° - 40° = 30°,$$

as claimed.   □

Notes. Edward Mann Langley (1851–1933) read for the Mathematical Tripos at Trinity College, Cambridge, and in 1878 he was Eleventh Wrangler. He is known for founding *The Mathematical Gazette* in 1894, and for posing the Adventitious Angles problem above. I first heard of this problem in the early 1970s from J.E. Littlewood (1885–1977), who considered it a lovely curiosity. I discovered only recently that the problem had been solved by James Mercer (1883–1932) in the way above (and the way I rediscovered fifty years ago).

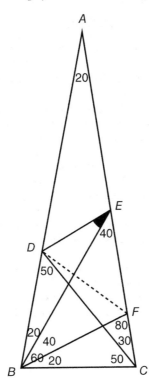

Figure 19 Adventitious angles

As in 1907 two Trinity men, Littlewood and Mercer, were bracketed Senior Wranglers, Littlewood always had a healthy respect for Mercer. It is not impossible that this was also the reason Littlewood remembered this problem and occasionally challenged people with it.

Adventitious angles, i.e. rational multiples of π, have also given rise to genuine mathematical research: see the references below. In particular, adventitious quadrangles have been studied by Bol, Kong, Zhang, Quadling, Rigby and Tripp, and the related problem of intersections of diagonals by Steinhaus, Croft, Fowler, Harborth, Rigby, Poonen and Rubenstein, to name some. For example, Steinhaus conjectured that for a prime p no three diagonals of a regular p-gon intersect in the same point: this was proved by Croft and Fowler. Although the proof of this lovely result is not very intricate, it is more complicated than most of the proofs in the present collection.

References

Bol, G., Beantwoording van prijsvraag no. 17, *Nieuw Archief voor Wiskunde* **18** (1930) 14–66.

Croft, H.T. and M. Fowler, On a problem of Steinhaus about polygons, *Proc. Camb. Phil. Soc.* **57** (1961) 686–688.

Harborth, H., Diagonalen im regularen *n*-Eck, *Elem. Math.* **24** (1969) 104–109.

Kong, Y. and S. Zhang, The adventitious angles problem: The lonely fractional derived angle, *Amer. Math. Monthly* **123** (2016) 814–816.

Langley, E.M., Problem 644, *Math. Gaz.* **11** (1922) 173.

Poonen, B. and M. Rubinstein, The number of intersection points made by the diagonals of a regular polygon, *SIAM J. Discrete Math.* **11** (1998) 135–156.

Quadling, D.A., The adventitious angles problem: A progress report, *Math. Gaz.* **61** (1977) 55–58.

Quadling, D.A., Last words on adventitious angles, *Math. Gaz.* **62** (1978) 174–183.

Rigby, J.R., Adventitious quadrangles: A geometrical approach, *Math. Gaz.* **62** (1978) 183–191.

Rigby, J.R., Multiple intersections of diagonals of regular polygons, and related topics, *Geom. Dedicata* **9** (1980) 207–238.

Steinhaus, H., Problem 225, *Colloq. Math.* **5** (1958) 235.

Tripp, C.E., Adventitious angles, *Math. Gaz.* **59** (1975) 98–106.

15. The Tantalus Problem – from *The Washington Post*

Let ABC be an isosceles triangle with angle 20° at the apex A, and so angles 80° at the base; furthermore, let D be a point on the side AB such that $\angle CD = 10°$, and E on AB such that $\angle BE = 20°$. By entirely elementary methods, without any recourse to trigonometry, show that $\angle DE = 20°$.

Proof.

As in the *Hint*, reflect A in CD to get a point H. Also, let G be on AB such that EG is parallel to BC. Then the triangles ABC and CAH are congruent, with angles 20°, 80° and 80°. Furthermore, $\angle CH = \angle CG = 60°$, so the points C, G and H are collinear. Add one more point, the intersection F of BE and CG: then the four points C, F, G and H are collinear. Since G is the intersection of AB and CH, and the triangles ABC and CHA are congruent, $BG = HG$. Next, note that the triangles BCF, EFG and ADH are equilateral, as all their angles are 60°; see the figure. Hence, $\angle DC = \frac{1}{2}\angle DH = 30°$.

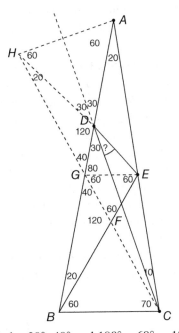

The triangles BGF and HGD have angles 20°, 40° and $180° - 60° = 120°$ and equal longest sides $BG = HG$, so they are congruent; *a fortiori*, $GF = GD$. Recalling that the triangle EGF is equilateral, we find that $GD = GE$. As the angle at the apex G of the isosceles triangle GDE is 80°, we have $\angle DE = 50°$.

Finally, $\angle DC = 30°$, so

$$\angle DE = \angle DE - \angle DC = 50° - 30° = 20°,$$

as claimed.   □

Notes. The alert (and perhaps even not-so-alert) reader must have noticed that the style of the previous problem and its present 'younger brother' is rather different from the style of the other problems in this collection. That one was included because of its connection with Langley, Mercer and Littlewood, and this one has been included for rather different reasons.

The problem is from the column of Carolyn Hax in the 13th September 1995 issue of *The Washington Post*. In this collection this problem appears as a nod to popular puzzles, and to emphasize that, as a standalone problem of no consequence, it is not the kind of problem that could ever appeal to real mathematicians. The problem is certainly non-trivial, and it is great that some readers of a popular newspaper tried to solve it. However, the remarks on it are shockingly misleading as to what mathematicians do and how good they are. Here is part of this commentary, as reproduced on p. 292 of Heilbron's book, and so taken at face value.

In 1995 The Washington Post *posed a geometrical puzzle that, for its apparent ease and real difficulty, might be called the Tantalus problem. Some readers who cracked their heads but not the problem complained that the puzzler had not supplied enough information for a solution. The United Press Syndicate, which had carried the quiz in which the puzzle first appeared, insisted it could be solved, but by a method too long to print. The man who set the puzzle, whose business is to write study guides for college aptitude tests, said that he had forgotten how to do it and could not repeat his lost performance. He had recourse to three dozen geometers, none of whom, he said, could find a solution. 'I contacted about 40 geniuses around the nation and they all gave me insights about the problem without being able to solve it.' With these insights and a weekend's labor he managed a solution.*

Reference

Heilbron, J.L., *Geometry Civilized – History, Culture and Technique*, Clarendon Press (1998).

16. Pythagorean Triples

A triple (a, b, c) of natural numbers is a primitive Pythagorean triple with a odd and b even if and only if there are relatively prime numbers $u > v \geq 1$ of opposite parity such that $a = u^2 - v^2$, $b = 2uv$ and $c = u^2 + v^2$.

First Proof   Let (a, b, c) be a primitive Pythagorean triple with a odd and b even. Then c is odd, so $a + c$ and $a - c$ are even. Hence

$$b^2 = c^2 - a^2 = (c + a)(c - a)$$

tells us that

$$\left(\frac{b}{2}\right)^2 = \left(\frac{c + a}{2}\right)\left(\frac{c - a}{2}\right) = xy,$$

say.

Are the integers x and y relatively prime? Yes, they are, since if $x = tx'$ and $y = ty'$ for some natural numbers $t \geq 2$, x' and y' then $c = x + y = t(x' + y')$ and $a = x - y = t(x' - y')$, implying that t is a common divisor of a and c, which is a contradiction. Now, since x and y are relatively prime and their product is a square (namely the square of the integer $(b/2)$), x and y are squares themselves: $x = u^2$ and $y = v^2$, say, where u and v are natural numbers. Therefore $a = u^2 - v^2$, $b = 2uv$ and $c = u^2 + v^2$, as claimed. Furthermore, as a and c are odd, u and v have opposite parities. Finally, since a and c are relatively prime, so are u^2 and v^2, and so u and v as well.

Conversely, if u and v are relatively prime natural numbers of opposite parities, with $u > v$, say, then $a = u^2 - v^2$, $b = 2uv$ and $c = u^2 + v^2$ form a relatively prime Pythagorean triple since a and c are relatively prime and

$$(2uv)^2 + (u^2 - v^2)^2 = (u^2 + v^2)^2.$$

This completes our proof.   □

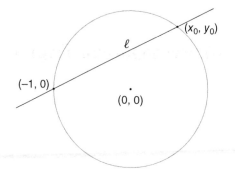

Figure 20 Here $x_0 = 3/5$, $y_0 = 4/5$ and the line ℓ is given by $y = \left(\frac{y_0}{x_0+1}\right)(x+1) = \frac{x+1}{2}$.

Second Proof Pythagorean triples (a, b, c) in their lowest form correspond to points (x_0, y_0) on the unit circle $x^2 + y^2 = 1$ with x_0 and y_0 strictly positive rational numbers. The line ℓ through such a point (x_0, y_0) and $(-1, 0)$ has equation

$$y = \alpha(x + 1),$$

with $\alpha = y_0/(x_0 + 1)$ rational, see Figure 20.

Hence to find all such rational points (x_0, y_0) it suffices to find the rational slopes α with $0 < \alpha < 1$ such that the line ℓ meets the unit circle $x^2 + y^2 = 1$ in $(-1, 0)$ *and* a rational point. Now, ℓ meets the unit circle in the points satisfying

$$y = \alpha(x + 1) \quad \text{and} \quad x^2 + y^2 = 1.$$

Solving this system of equations, we find that

$$(1 + \alpha^2)x^2 + 2\alpha^2 x + (\alpha^2 - 1) = 0.$$

Since $x = -1$ is a root, $(1 + x)$ is a factor of the left-hand side:

$$(1 + x)\left((1 + \alpha^2)x + (\alpha^2 - 1)\right) = 0,$$

so $x = (1 - \alpha^2)/(1 + \alpha^2)$ and $y = 2\alpha/(1 + \alpha^2)$. Writing the rational α in its lowest form, so that $\alpha = v/u$ for two relatively prime positive integers $u > v$, we find that

$$x = \frac{u^2 - v^2}{u^2 + v^2} \quad \text{and} \quad y = \frac{2uv}{u^2 + v^2}.$$

Therefore $a = u^2 - v^2$, $b = 2uv$ and $c = u^2 + v^2$ give us all relatively prime Pythagorean triples (a, b, c). $\square$

Notes. Saying that this is an old chestnut does not do justice to the reputation of this simple characterization of the Pythagorean triples. Still, I have put it into this collection because I doubt that many people encounter it in school, although I am convinced they should. I remember that I was very impressed by this characterization when in my early teens (or just before) I first came across it. It seemed magical that one could construct *all* Pythagorean triples in such a simple way. Amazingly, I don't think that this beautiful little result was ever mentioned to us in school, although I may be unfair to my excellent teachers, and I just don't remember the occasion.

17. Fermat's Theorem for Fourth Powers

The equation $a^4 + b^4 = c^4$ has no solutions in natural numbers.

Proof. Let us prove more, namely that the equation

$$a^4 + b^4 = c^2$$

has no solutions in positive integers. Suppose that this assertion is false, and let (a, b, c) be its solution with c *minimal.* Our aim is to arrive at a contradiction.

We may and shall assume that a, b and c (or just any two of them) are relatively prime, and then that the notation is chosen so that b is even, and a and c are odd. Since

$$(a^2)^2 + (b^2)^2 = c^2,$$

(a^2, b^2, c) is a Pythagorean triple, so by the characterization in the preceding problem, there are relatively prime positive integers $u > v$ such that they are of opposite parity,

$$a^2 = u^2 - v^2, \quad b^2 = 2uv \quad \text{and} \quad c = u^2 + v^2.$$

The first of these equations is again a Pythagorean equation in relatively prime numbers:

$$a^2 + v^2 = u^2,$$

so u is odd and v is even. By the characterization of relatively prime Pythagorean triples, there are relatively prime positive integers $u_1 > v_1$ such that

$$a = u_1^2 - v_1^2, \quad v = 2u_1v_1 \quad \text{and} \quad u = u_1^2 + v_1^2.$$

To complete our solution, we shall show that the third of these equations is of the same type as the original equation $a^4 + b^4 = c^2$, i.e. each of u, u_1 and v_1 is itself a square; as $u < \sqrt{c}$, this will contradict the minimality of c.

84

Since u and v are relatively prime, u is odd and $b^2 = 2uv$, we find that u and $2v$ are squares: $u = x^2$ and $2v = (2y)^2$. But then

$$2y^2 = v = 2u_1v_1;$$

as u_1 and v_1 are relatively prime and their product is the square y^2, both of them are squares:

$$u_1 = a_1^2 \quad \text{and} \quad v_1 = b_1^2.$$

Finally, with $c_1 = x$ we find that $u = c_1^2$, so

$$a_1^4 + b_1^4 = c_1^2.$$

Since $c_1 = x < c$, this contradicts the minimality of c, and so our proof is complete. $\square$

Notes. As no doubt every reader has realized, this exercise is the special case $n = 4$ of Fermat's Last Theorem that for $n \geq 3$ the equation $a^n + b^n = c^n$ has no solutions in strictly positive integers. Not only is $n = 4$ a special case, but it is also the easiest special case, already proved by Fermat himself in 1637: it is a natural rider to the characterization of Pythagorean triples in the preceding problem. I should like to emphasize that this case has *nothing, absolutely nothing* to do with the phenomenal proof Andrew Wiles gave of Fermat's Last Theorem in 1994, making essential use of the results in his great paper with Richard Taylor, published right after Wiles's masterpiece. The importance of these results is just about impossible to overestimate.

References

Taylor, R. and A. Wiles, Ring-theoretic properties of certain Hecke algebras, *Ann. of Math. (2)* **141** (1995) 553–572.

Wiles, A., Modular elliptic curves and Fermat's last theorem, *Ann. of Math. (2)* **141** (1995) 443–551.

18. Congruent Numbers – Fermat

The natural number 1 *is not a congruent number, i.e. there is no right-angled triangle with rational sides, whose area is* 1.

Proof. Let us suppose that 1 is a congruent number, i.e. there are rational numbers $r_1 = p_1/q_1$, $r_2 = p_2/q_2$ and $r_3 = p_3/q_3$, with p_i and q_i positive integers, such that $r_1^2 + r_2^2 = r_3^2$ and $r_1 r_2/2 = 1$. Then, with $q = q_1 q_2 q_3$, the triangle with integer sides $a = qr_1$, $b = qr_2$ and $c = qr_3$ is right-angled, and its area is a square, q^2. Putting it another way, there is a Pythagorean triple (a, b, c) such that the area $ab/2$ of the corresponding triangle is a square. Our task is to show that this is impossible. We prove this by the method of infinite descent (to call a triviality a 'method'): we take such a triple (a, b, c) with c minimal, and prove that there is another Pythagorean triple (a', b', c') such that $a'b'/2$ is a square and $c' < c$.

Since c is minimal, the Pythagorean triple (a, b, c) is primitive so, recalling Problem 16, we may assume that $a = n^2 - m^2$, $b = 2mn$ and $c = m^2 + n^2$, where m and n are relatively prime, with one even and the other odd.

The area A of this triangle is $A = ab/2 = nm(n + m)(n - m)$, with the four factors relatively prime. Since A is a square, each of the four factors is a square: say, $n = x^2$, $m = y^2$, $n + m = s^2$ and $n - m = t^2$, with x, y, s and t relatively prime natural numbers, and both s and t odd. We shall construct a suitable Pythagorean triple (a', b', c') with $c' = x = \sqrt{n} < n^2 + m^2 = c$. First, note that

$$2x^2 = 2n = s^2 + t^2, \tag{1}$$

and

$$2y^2 = 2m = s^2 - t^2 = (s + t)(s - t). \tag{2}$$

Since s and t are relatively prime odd numbers, if d divides both $s + t$ and $s - t$, then it divides $2s = (s + t) + (s - t)$ and $2t = (s + t) - (s - t)$ as well,

86

implying that it divides 2. Hence the greatest common divisor of $s + t$ and $s - t$ is 2. Consequently, the two factors in (2) are of the form $2q^2$ and $4r^2$ in one of the two possible ways, with q odd. This implies that $2y^2 = 8(qr)^2$, so $y = 2qr$, and also that

$$s = \frac{1}{2}\left((s + t) + (s - t)\right) = q^2 + 2r^2,$$

and

$$t = \frac{1}{2}\left((s + t) - (s - t)\right) = \frac{1}{2}|2q^2 - 4r^2| = |q^2 - 2r^2|.$$

Recalling (1), this tells us that

$$x^2 = n = \frac{1}{2}(s^2 + t^2) = \frac{1}{2}\left((q^2 + 2r^2)^2 + (q^2 - 2r^2)^2\right) = q^4 + 4r^4,$$

so $(q^2, 2r^2, x)$ is a Pythagorean triple. The area of the right-angled triangle with these sides is the square $2q^2r^2/2 = (qr)^2$. Finally, we have already noted that the hypotenuse of this triangle, x, is much smaller than $c = n^2 + m^2$.   □

Notes.

The result above was proved by Pierre de Fermat (1607–1665), the great French mathematician. He is best known for 'Fermat's Little Theorem' and even more for 'Fermat's Last Theorem', whose statement jotted down in the margin of his copy of Diophantus's *Arithmetica*, which was proved 358 years later by Andrew Wiles.

The beautiful proof of the fact that 1 is not a congruent number we have just presented is Fermat's: we followed John Coates's presentation of this proof in 2014.

Coates wrote in 2014 and 2017 that 'the oldest unsolved major problem in number theory, and possibly in the whole of mathematics' is the congruent number problem: that of deciding which positive integers are congruent numbers. In spite of this undeniable claim to fame, this problem has received much less attention than Fermat's Last Theorem, whose proof in 1995 by Andrew Wiles, aided by Richard Taylor, was a mathematical sensation of the 20th century.

In the 10th century, and possibly earlier, Arab (and probably Indian) mathematicians found that the numbers

$$5, 6, 7, 13, 14, 15, 21, 22, 23, 29, 30, 31, 34, 37, 38, 39, 41, 46, 47, \ldots$$

are all congruent. Some of these numbers are easily seen to be congruent, while others need a fair amount of work.

Fermat's result above has some connection with his Last Theorem: it implies that the equation $x^4 - y^4 = z^2$ has no solutions in non-zero integers. Indeed, suppose that this equation does have a solution in positive integers. Assuming, as we may, that $x > y$, set $n = x^2$, $m = y^2$, take the Pythagorean triple n and m and define: $a = n^2 - m^2$, $b = 2nm$ and $c = n^2 + m^2$. The area of this Pythagorean triangle is $(n^2 - m^2)nm = x^2 y^2 z^2$. Dividing the length of each side by xyz, we get a right-angled triangle with rational sides and area 1, contradicting Fermat's result above.

Needless to say, if $x^4 - y^4 = z^2$ has no solutions in non-zero integers, neither does $x^4 + y^4 = z^4$, so this is another solution of the previous problem. Over the centuries, a host of other proofs have been found for this. A consequence of this simple result is that, in proving that for $n \geq 3$ the equation $x^n + y^n = z^n$ has no solution in positive integers, we may assume that n is a prime.

Finally, let us note that putting together the results in this problem and in the previous one, we find that if in the equation $x^{n_1} + y^{n_2} = z^{n_3}$ two of the exponents are 4 and the third is 2, then the equation has no solution in natural numbers. However, if two of the exponents are 2 and the third is 4, then there are solutions, e.g. $40^2 + 3^4 = 41^2$ and $24^2 + 7^2 = 5^4$.

References

Coates, J., Congruent numbers *Acta Math. Vietnam.* **39** (2014) 3–10.

Coates, J., The oldest problem, *ICCM Not.* **5** (2017) 8–13.

Taylor, R. and A. Wiles, Ring-theoretic properties of certain Hecke algebras, *Ann. of Math. (2)* **141** (1995) 553–572.

Wiles, A., Modular elliptic curves and Fermat's last theorem, *Ann. of Math. (2)* **141** (1995) 443–551.

19. A Rational Sum

A necessary and sufficient condition on a rational number $s > 1$ *such that* $\sqrt{s+1} - \sqrt{s-1}$ *is also rational is that* $s = (c^4 + 4d^4)/(4c^2d^2)$ *for two integers* c *and* d.

Proof. Suppose that $s = a/b$ for two relatively prime integers a and b, $a > b$, and $\sqrt{s+1} - \sqrt{s-1}$ is a rational number. Squaring $\sqrt{s+1} - \sqrt{s-1}$, we find that $\sqrt{s^2 - 1}$ is also rational, say

$$s^2 - 1 = u^2/v^2,$$

where u and v are relatively prime natural numbers. Hence,

$$(a^2 - b^2)v^2 = b^2u^2.$$

Since u^2 and v^2 are relatively prime, and so are b^2 and $a^2 - b^2$, we find that $b = v$, and so

$$a^2 = b^2 + u^2.$$

By the characterization of Pythagorean triples, one of two cases holds:

(i) $a = m^2 + n^2$, $b = m^2 - n^2$, $u = 2mn$,
(ii) $a = m^2 + n^2$, $u = m^2 - n^2$, $b = 2mn$,

where $m > n$ are relatively prime natural numbers, with $m + n$ odd.

Case (i) leads to the contradiction that the following expression should be rational:

$$\left(\frac{m^2 + n^2}{m^2 - n^2} + 1\right)^{1/2} + \left(\frac{m^2 + n^2}{m^2 - n^2} - 1\right)^{1/2} = \frac{\sqrt{2}(m + n)}{\sqrt{m^2 - n^2}} = \sqrt{2}\sqrt{\frac{m + n}{m - n}}.$$

Since $m + n$ and $m - n$ are odd, this clearly does not hold. Hence case (ii) holds, so $a = m^2 + n^2$ and $b = m^2 - n^2$.

As the expression

$$\left(\frac{m^2 + n^2}{2mn} + 1\right)^{1/2} + \left(\frac{m^2 + n^2}{2mn} - 1\right)^{1/2} = \frac{2m}{\sqrt{2mn}}$$

is rational, we have $2m/n = c^2/d^2$ for some relatively prime natural numbers c and d. If n is even, this gives $m = c^2$ and $n = 2d^2$, so

$$a = c^4 + 4d^4 \quad \text{and} \quad b = 4c^2d^2. \tag{1}$$

If m is even, $m/2 = (c/2)^2$ and $n = d^2$, where c is even, so

$$a = 4(c/2)^4 + d^4 \quad \text{and} \quad b = 4(c/2)^2d^2.$$

This is precisely (1) with the notation changed, so we may assume that (1) holds. Hence $s = (c^4 + 4d^4)/(4c^2d^2)$, proving the necessity of this condition.

The sufficiency is trivial: if $s = (c^4 + 4d^4)/(4c^2d^2)$ then

$$\sqrt{s+1} - \sqrt{s-1} = \frac{c^2 + 2d^2}{2cd} - \frac{c^2 - 2d^2}{2cd} = \frac{2d}{c}$$

is indeed rational.   □

Notes. This exercise is one of the more pedestrian and simple examples used by John Hammersley of Trinity College, Oxford, in his paper expounding his legendary lecture "On the enfeeblement of mathematical skills by 'Modern Mathematics' and by similar soft intellectual trash in schools and universities"

he delivered at the Annual General Meeting of the Institute of Mathematics and its Applications on 8th June 1967. Let me quote the stunning start of this article.

The word 'modern' comes from the Latin modo, *meaning just now, here today, gone tomorrow, ephemeral. More often than not we speak in this sense of modern art. Here is the artist setting aside established traditional skills and having instead his contemporary fling: put the canvas on the floor and paint on your bicycle wheels, ride around and hope that someone else's psyche will make something of the result. From all such entirely legitimate experiment only a tiny fraction will survive as an original and significant addition to human achievement, and the rest will be forgotten as all yesterday's nonentities and trivia have always been. Collectors know how hard it is to spot tomorrow's gems in today's midden. On the other hand, speaking of modern languages we mean as a rule living languages, words and expressions which men use in their daily affairs for their current thoughts and emotions, for chatter, for diplomacy, for professional and technical communication, and for commercial traffickings. That contrasts a modern with a dead language. The latter has its splendours and is not obsolete; but it is out of date and, as any encaenia shows only too plainly, it lacks the terminology and range of expression to handle many of the concepts that mankind has now come to hold important or to reach towards. I want to consider later whether modern mathematics is modern in the sense of modern art or modern languages.*

To conclude his paper, Hammersley gives sixteen problems of greatly varying difficulties that school students and undergraduates ought to be able to tackle. The problem in this exercise is ninth in the list, and is one of the easiest.

Reference

Hammersley, J., On the enfeeblement of mathematical skills by 'Modern Mathematics' and by similar soft intellectual trash in schools and universities, *Bull. Inst. Math. Appl.* **4** (1968) 66–85.

20. A Quartic Equation

Find a large family of integer solutions of

$$A^4 + B^4 = C^4 + D^4. \tag{1}$$

More precisely, look for fairly general polynomials A, B, C and D in $\mathbb{Z}[a,b]$ such that (1) holds. To this end, look for the solution in the form

$$A = ax + c, \quad B = bx - d, \quad C = ax + d, \quad D = bx + c \tag{2}$$

where a, b, c, d and x are rational numbers. Considering a, b, c and d constant, (1) holds if x satisfies a quartic whose first and last coefficients are 0. Show that with a suitable choice of a, b, c and d the coefficient of x^3 is also 0, and use this to find our polynomials.

Proof. With $A = ax + c$, $B = bx - d$, $C = ax + d$ and $D = bx + c$, as above, (1) holds if

$$4(a^3c - b^3d - a^3d - b^3c)x^3 + 6(a^2c^2 + b^2d^2 - a^2d^2 - b^2c^2)x^2$$
$$+ 4(ac^3 - bd^3 - ad^3 - bc^3)x = 0.$$

Hence, the coefficient of x^3 is 0 if

$$c(a^3 - b^3) = d(a^3 + b^3).$$

This certainly holds if $c = a^3 + b^3$ and $d = a^3 - b^3$: this is what we shall take. Dividing our equation by $2x$, we get

$$3x(a^2c^2 + b^2d^2 - a^2d^2 - b^2c^2) = 2(-ac^3 + bd^3 + ad^3 + bc^3),$$

i.e.

$$3x(a^2 - b^2)(c^2 - d^2) = 2c^3(b - a) + 2d^3(a + b).$$

92

Substituting $c = a^3 + b^3$ and $d = a^3 - b^3$ we find that

$$12xa^3b^3(a^2 - b^2) = 4ab(a^4 - b^4)(a^4 - 3a^2b^2 + b^4),$$

so (1) certainly holds if

$$3a^2b^2x = (a^2 + b^2)(a^4 - 3a^2b^2 + b^4).$$

Finally, substituting for x, c and d in (2) and multiplying throughout by $3a^2b^2$, we arrive at the polynomials

$$A = a^7 + a^5b^2 - 2a^3b^4 + 3a^2b^5 + ab^6,$$
$$B = a^6b - 3a^5b^2 - 2a^4b^3 + a^2b^5 + b^7,$$
$$C = a^7 + a^5b^2 - 2a^3b^4 - 3a^2b^5 + ab^6,$$
$$D = a^6b + 3a^5b^2 - 2a^4b^3 + a^2b^5 + b^7.$$

These polynomials satisfy the quartic equation (1), as required. □

Notes. Leonhard Euler (1707–1783) was the first to give integral solutions of the equation (1): in 1772 he gave homogeneous polynomials of degree seven, with integral coefficients, in two parameters f and g for A, B, C and D. (See Dickson (1929), pp. 60–62, and Hardy and Wright (2008), p. 202.) In 1915 Gérardin gave a simpler solution, which he first published as a problem; a little later Rignaud pointed out that this simpler solution can be obtained from Euler's by the substitutions $f = a + b$ and $g = a - b$.

This exercise is the simple and intuitive proof of Gérardin's result that Sir Peter Swinnerton-Dyer (1927–2018) gave when he was a scholar at Eton, just after his 15th birthday. At Eton Sir Peter wasn't yet the 16th Baronet that he later became, and the paper is signed as P.S. Dyer; Louis Mordell (1888–1972) kindly provided the young schoolboy with appropriate references. At the time Mordell was already a famous professor in Manchester: in fact, in the 1930s he built up a tremendous group of mathematicians there, including Kurt Mahler (1903–1988), Harold Davenport (1907–1969), Chao Ko (1910–2002) and Paul Erdős (1913–1996). Later, when Hardy retired, Mordell returned to Cambridge, where he remained a Fellow of St John's till the end of his life. He often joked that the Americans had sent him to Cambridge from Philadelphia to be the last First Wrangler, but he blotted his copybook, and was only Third Wrangler.

In his brief paper, Swinnerton-Dyer notes that, using his method above, from

$$p^4 + q^4 = r^4 + s^4$$

further solutions can be obtained. Indeed, set

$$A = ax + p, \quad B = bx + q, \quad C = ax + r, \quad D = bx + s,$$

and choose a and b so as to make the coefficient of x in the resulting equation zero. An obvious solution for x is

$$x = \frac{3a^2(r^2 - p^2) - 3b^2(q^2 - s^2)}{2a^3(p - r) + 2b^3(q - s)},$$

which gives new values for A, B, C and D.

After Eton, Swinnerton-Dyer went up to Trinity College, Cambridge to read for the Mathematical Tripos, and got a Fellowship before finishing his PhD. For decades, he was a mainstay of Trinity College, and one of the most influential men in academic politics in the UK. In mathematics, he is best known for the Birch–Swinnerton-Dyer Conjecture in algebraic geometry, which is one of the most important problems in mathematics, and one of the seven million-dollar Millennium Problems.

References

Dickson, L.E., *Introduction to the Theory of Numbers*, The University of Chicago Press, (1929).

Dyer, P.S., A solution of $A^4 + B^4 = C^4 + D^4$, *J. London Math. Soc.* **18** (1943) 2–4.

Gérardin, A., *Intermédiaire des Math.* **24** (1917) 51.

Hardy, G.H. and E.M. Wright, *An Introduction to the Theory of Numbers*, Sixth edition, Revised by D.R. Heath-Brown and J.H. Silverman, with a foreword by Andrew Wiles, Oxford University Press (2008).

Rignaud, A., *Intermédiaire des Math.* **25** (1918) 27–28; 133–134.

21. Regular Polygons

Of all polygons with the same number of sides and equal perimeter length, the regular polygon has the greatest area.

Proof. For $n \geq 3$, write a_n for the area of the regular n-gon with side length 1. We shall show that every n-gon with perimeter length *at most* n has area *at most* a_n. This is trivially equivalent to the statement that every n-gon with perimeter length *at most* n has area *at most* a_n.

We apply induction on n. We can start with the trivial base case $n = 3$, although even that will be proved below.

For $n \geq 3$, let P_n be a polygon of perimeter n with the greatest area $m_n \geq a_n$. (That there is such a polygon P_n is a simple consequence of the compactness of the set of such polygons with the Hausdorff distance, but we shall not go into this – compactness was not a friend of the ancient Greeks, not even of the great geometer Jakob Steiner in the 19th century.) Clearly, P_n is also a polygon with minimal perimeter length among all n-gons with area at least m_n. Our task is to show that P_n is equilateral and equiangular. We shall accomplish this in several steps.

We shall write XY for the segment joining a point X to a point Y, and also its length. Also, we write $\lambda(P)$ for the area ('Lebesgue measure') of a polygon P, so that $\lambda(ABC)$ is the area of the triangle ABC and $\lambda(ABCD)$ is the area of the quadrilateral $ABCD$.

(i) *P_n is equilateral.*

To see this, let A, B and C be three consecutive vertices of P_n, and suppose that, with the obvious notation, $AB \neq BC$. We have to show that this leads to a contradiction.

Reflect B in the perpendicular bisector of AC to obtain B' and let B^* be the midpoint of the segment BB', as in Figure 21. Then the triangles ABC and AB^*C have the same area, and the points B and B' are on the ellipse consisting

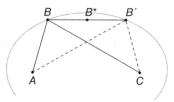

Figure 21 The points B, B' on the ellipse $AX + XC = AB + BC$, and the midpoint B^* of the segment BB'.

of the points X with $AX + XC = AB + BC$. Since an ellipse is convex, the midpoint B^* is in the interior of our ellipse, so $AB^* + B^*C < AB + BC$, contradicting the definition of P_n.

(ii) *The assertion holds for* $n = 3$ *and* 4.

For $n = 3$ we are done by (i): our triangle P_3 is equilateral, so regular. Also, for $n = 4$, part (i) tells us that P_4 is a rhombus. Since the area of a rhombus with each side having length 1 has area at most 1, the area of a unit square, so we are again done. From now on, we may assume that $n \geq 5$.

(iii) *Let* A, B, C *and* D *be four consecutive vertices of* P_n, *where* $n \geq 5$, *so that* $AB = BC = CD = 1$. *Then* $AD > 1$.

Indeed, otherwise, by the induction hypothesis, the quadrilateral $ABCD$ has area at most a_4, and the remaining $(n - 2)$-gon has area at most a_{n-2}, so $m_n \leq a_4 + a_{n-2} < a_n \leq m_n$, implying our assertion.

(iv) *The polygon* P_n *is equiangular.*

This assertion is the heart of the proof: once it is proved, our assertion is proved as well. To show (iv), it suffices to prove that if A, B, C and D are four consecutive vertices of P_n then the angles at B and C are equal, i.e. the side BC is parallel to the chord AD. We suppose that this is not the case, so that our aim is to arrive at a contradiction. We may assume that B is further from the line AD than C.

We start with some notation. Reflect B in the perpendicular bisector of the segment AD to obtain B'; similarly, let C' be the reflection of C. Let E be the midpoint of the segment BC', and F the midpoint of the segment $B'C$. Let B^* be the projection of B into the line AD, and define C^*, E^* and F^* analogously, so that $b = BB^*$ is the distance of B from the line AD and $c = CC^* < b$ is the distance of C from the line AD. Write d for the length of the diagonal AD, so that $d > 1$. Since $1 = BC < d = AD$, the angle ADC is less than $\pi/2$. We shall distinguish two cases according to the size of the angle DAB.

(a) *The angle* DAB *is at most* $\pi/2$, *i.e. the points* B^* *and* C^* *are on the segment*

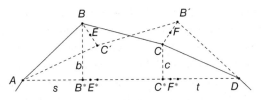

Figure 22 The sides have length 1: $AB = BC = CD = AC' = C'B = B'D = 1$;
also, $AB^* = s, BB^* = b, CC^* = c, DC^* = t$, etc.

AD. Denote the (length of the) segment AB^* by s and DC^* by t, so that $s < t$.
Note that the projection of B' is at distance s from D and the projection of C'
is at distance t from A, with all these projections on the segment AD. Also,
$AE^* = (s + t)/2 = DF^*$, as in Figure 22.

Then each of the segments AB, DC, AC' (the reflection of DC), DB' (the
reflection of AB), BC and $B'C'$ (the reflection of BC) has length 1, so the
segments AE and DF also have length less than 1. Finally, $EF = E^*F^* =
d - s - t = B^*C^* < BC = 1$. The contradiction we seek is that the area of
the quadrilateral $AEFD$ is greater than the area of the original quadrilateral
$ABCD$. This is easily done by reading off the areas of various quadrilaterals
and triangles. Indeed, set

$$S = \lambda(B^*BCC^*) = (d - s - t)(b + c)/2 = \lambda(E^*EFF^*),$$

so that

$$\lambda(ABCD) = \lambda(ABB^*) + \lambda(B^*BCC^*) + \lambda(C^*CD) = bs/2 + S + ct/2,$$

and

$$\lambda(AEFD) = \lambda(AEE^*) + \lambda(E^*EFF^*) + \lambda(F^*FD) = (b + c)(s + t)/4 + S.$$

Consequently,

$$\lambda(AEFD) - \lambda(ABCD) = (b + c)(s + t)/4 - bs/2 - ct/2 = (b - c)(t - s)/4 > 0,$$

as claimed. This proves the result in case (a).

(b) *The angle DAB is greater than $\pi/2$, so on the line AD our points are in this
order:* B^*, A, E^*, C^*, F^*, D. The calculations remain exactly the same as in case
(a), except that in the formulae above $s < t$ has to be replaced by $-s$. Then we
get $\lambda(AEFD) - \lambda(ABCD) = (b + c)(t - s)/4 + bs/2 - ct/2 = (b - c)(s + t)/4 > 0$,
proving the case (b), and so telling us that P_n is indeed equiangular.

Let us repeat the obvious conclusion: as by (i) the polygon P_n is equilateral,
and by (iv) it is equiangular, it is indeed regular, completing the proof of our
exercise.   □

Notes. The result above is one of several theorems of Zenodorus ($Z\eta\nu o\delta\omega\rho o\zeta$) from his little treatise *On Isoperimetric Figures*. Although this treatise is now lost, we know about it from later commentaries. Zenodorus was a Greek mathematician, who lived from about 200BC till about 140BC: he was the first to study isoperimetric problems, i.e. the connection between the area and perimeter of a domain. As we know from Sir Thomas Heath's *The History of Greek Mathematics*, in about 130BC 'Polybius observed that there were people who could not understand that camps of the same periphery might have different capacities'. Among other results, Zenodorus proved (not entirely rigorously) that a circle is greater than any regular polygon of equal contour, and of two regular polygons of equal perimeter, the one with more angles has the greater area. Note that the latter assertion is immediate from the result in this problem above.

Sir Thomas Heath read mathematics *and* classics at Trinity College, Cambridge, gaining a first class degree in both. He finished Twelfth Wrangler in 1882, and gained a Fellowship of Trinity College: years later he became an Honorary Fellow. He was a most distinguished civil servant and one of the greatest experts on Greek mathematics, translating several of the outstanding mathematicians.

The proof we have given is not the simplest (far from it!), but it is one of the 'pedestrian' proofs that are easy to find. Later we shall give two simpler proofs of stronger assertions, but those simple and elegant proofs require unexpected ideas.

Reference

Heath, Sir Thomas, *A History of Greek Mathematics: Vol. II, From Aristarchus to Diophantus*, Clarendon Press (1921).

22. Flexible Polygons

The maximum of polygons, with all the sides given but one, may have a circle circumscribed about it, having the unknown side for a diameter of the circle.

Proof. Let $P_n = ABC \cdots Z$ be a polygon of maximum area formed by the given sides $AB, BC, \ldots, YZ$, and the unknown side, ZA. Clearly, P_n is convex. All we have to show is that if E is any vertex of P_n other than A and Z, then the angle AEZ is $\pi/2$.

Suppose this is not the case. Then, keeping the polygons $AB \cdots DE$ and $EF \cdots YZ$ rigid, rotate the chords (or sides) AE and EZ about E to make them perpendicular. This increases the area of the triangle AEZ, so increases the area of P_n, completing our proof. □

Notes. Strictly speaking, the argument above is incomplete: when we rotate the chords AE and EZ to make them perpendicular, we are in trouble if the rotated images of the polygons $AB \cdots DE$ and $EF \cdots YZ$ intersect. But that cannot happen, because the original angle DEF was less than π and we increase it by at most $\pi/2$.

For additional information, see the notes at the end of the next problem.

23. Polygons of Maximal Area

The area of a polygon with given sides $a_1, \ldots, a_n$ is not larger than the cyclic polygon with these sides, i.e. the one that may have a circle circumscribed about it.

First Proof   For $n = 3$ there is nothing to prove, so suppose that $n \geq 4$. Let $P_n^* = A^* B^* \cdots Y^* Z^*$ be the cyclic polygon with sides $a_1, \ldots, a_n$ – it is clear that P_n^* exists and is unique up to congruence. Let M^* be the point diametrically opposite A^*.

First we consider the case when M^* is not a vertex of P_n^*, and the cyclic order of our $n + 1$ points is $A^*, B^*, \ldots, E^*, M^*, F^*, \ldots, Y^*, Z^*$, where A^*, E^* and M^* are different and so are M^*, F^* and Z^*, as in Figure 23.

Now, given a polygon $P_n = AB \cdots YZ$ with sides $a_1, \ldots, a_n$, add to it a point M such that the triangle EFM is congruent to $E^* F^* M^*$. What can we say about the areas $\lambda(\cdots)$ of the various polygons? By Problem 21,

$$\lambda(AB \cdots EM) \leq \lambda(A^* B^* \cdots E^* M^*)$$

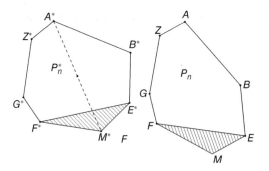

Figure 23 The divisions of the $(n + 1)$-gons obtained from P_n^* and P_n.

and

$$\lambda(MF \cdots YZA) \leq \lambda(M^*F^* \cdots Y^*Z^*A^*).$$

Consequently the area of the $(n + 1)$-gon $Q_{n+1} = AB \cdots EMF \cdots YZ$ is at most that of $Q^*_{n+1} = A^*B^* \cdots E^*M^*F^* \cdots Y^*Z^*$, and so the same holds for the n-gons P_n and P^*_n, whose areas are $\lambda(EMF)$ smaller.

The case when M^* is a vertex of P^*_n is even simpler. If M^* is preceded by E^* and followed by F^* then, again by Problem 21,

$$\lambda(AB \cdots EM) \leq \lambda(A^*B^* \cdots E^*M^*)$$

and

$$\lambda(MF \cdots YZA) \leq \lambda(M^*F^* \cdots Y^*Z^*A^*),$$

so

$$\lambda(P_n) \leq \lambda(P^*_n),$$

completing our first proof. □

Second Proof This proof is even more elegant than the first, but it relies on the fundamental result that the circle is the solution of the isoperimetric problem in the plane.

Let P_n be an n-gon with given sides, and P_{n^*} the cyclic n-gon with the same sides, inscribed in a circular disc of radius r. Attach each circular segment determined by a side K^*L^* of P^*_n to the side KL of P_n, and let D be the union of P_n and these n segments. (Our notation should not mislead the unsuspecting reader: we do *not* assume or claim that D is a circular disc – far from it.) Then D has the same perimeter length as D^*, so its area is less than that of D^*, unless D is also a circular disc of radius r. But then the area of P_n is less than that of P^*_n, unless it is inscribed in a circle. (Clearly, this circle has to have radius r.) This completes our second proof. □

Third Proof There are various ways of proving this basic result by the use of formulae. Here we deduce it from Bretschneider's formula from 1842, which used to feature in all books on plane geometry (like Hobson's 1918 book). Let P_n be a polygon with sides $a_1, \ldots, a_n$ and maximal area. The existence of such a polygon follows from a simple compactness argument, but here we gloss over it.

Since a circle is determined by three of its points, it suffices to show that any four consecutive vertices of P_n are on a circle. To this end, let a, b and c be the lengths of three consecutive sides, and let d be the length of the chord completing the quadrilateral. The existence of P_n tells us that among all quadrilaterals with sides a, b, c and d the one appearing in P_n has maximal area. Hence, all we need is that among all quadrilaterals with sides a, b, c and d the cyclic quadrilateral has maximal area. By Bretschneider's formula, a quadrilateral with sides a, b, c and d and two opposite angles α and γ, say, has area

$$\sqrt{(s-a)(s-b)(s-c)(s-d) - abcd\cos^2((\alpha+\gamma)/2)},$$

where s is the semiperimeter of the quadrilateral. Keeping a, b, c and d fixed, the maximum of this expression is attained when $\alpha + \gamma = \pi$, i.e. our quadrilateral is cyclic.   □

Notes. The first proof above is from Thomas Hill, see p. 48 of his 1863 book, and the second from the much more recent book of I.M. Yaglom and V.G. Boltyanskiĭ, and Bretschneider's formula, together with the obvious consequence that among all quadrilaterals with given sides the cyclic quadrilateral has maximal area, is from the 1918 book of E.W. Hobson (see pp. 204–205). Hobson must have discovered the formula for himself (not a great achievement!) because he does not mention Bretschneider, but attributes the special case of a cyclic polygon to 'Brahmegupta, a Hindoo Mathematician of the sixth century'. For the sake of completeness, we reproduce Hobson's proof.

Let $ABCD$ be a convex quadrilateral with sides $a = AB, b = BC, c = CD$ and $d = DA$, diagonals x and y, angle α at A and γ at C, as in Figure 24. Set $\xi = (\alpha + \gamma)/2$, write $s = (a + b + c + d)/2$ for the semiperimeter and S for the area of this quadrilateral.

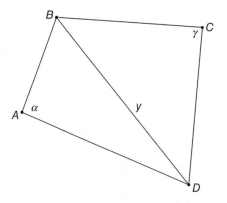

Figure 24 The notation for a convex quadrilateral.

Let us focus on the two triangles into which the diagonal BD of length y

divides our quadrilateral. First,

$$y^2 = a^2 + d^2 - 2ad \cos \alpha = b^2 + c^2 - 2bc \cos \gamma,$$

so we have

$$ad \cos \alpha - bc \cos \gamma = \frac{1}{2}(a^2 + d^2 - b^2 - c^2).$$

Next, taking the areas of these two triangles, we find that

$$ad \sin \alpha + bc \sin \gamma = 2S.$$

Squaring and then adding the last two equations, we find that

$$a^2 d^2 + b^2 c^2 - 2abcd \cos 2\xi = 4S^2 + \frac{1}{4}(a^2 + d^2 - b^2 - c^2)^2,$$

and so

$$16S^2 = 4(ad + bc)^2 - (a^2 + d^2 - b^2 - c^2)^2 - 16abcd \cos^2 \xi,$$

i.e.

$$16S^2 = \{(a + d)^2 - (b - c)^2\}\{(b + c)^2 - (a - d)^2\} - 16abcd \cos^2 \xi.$$

This can be rewritten in the form we have wanted to find Bretschneider's formula:

$$S^2 = (s - a)(s - b)(s - c)(s - d) - abcd \cos^2 \xi.$$

The four authors of the three books we have mentioned were all interesting mathematicians.

Thomas Hill (1818–1891) was not only a mathematician, but also (in fact, mostly) a clergyman, scientist, philosopher and inventor. He devised scientific instruments, and patented the *Hill Arithmometer*, an early key-driven adding machine. He was the 20th President of Harvard University.

Ernest William Hobson (1856–1933) was an English geometer and analyst: for most of his life he was at Christ's College, Cambridge, first as an under-graduate, and then as a Fellow. In the Mathematical Tripos in 1878 he was ranked Senior Wrangler. Through his important books, he did much to close the gap between British and Continental mathematics. He had two exceptional students, Philippa Fawcett and John Maynard Keynes. Fawcett, the alumna of Newnham College, Cambridge, shot to fame when she was 'ranked above the Senior Wrangler' in the Mathematical Tripos; Keynes of King's College, later The Lord Keynes, was perhaps the most influential economist of the 20th century.

Isaak Moiseevich Yaglom (1921–1988) was a Jewish–Russian mathematician of very broad interests and the author of several mathematics books for students, often written with his twin brother, Akiva. Finally, Vladimir Grigorevich Boltyanskiĭ (1925–2019) was an outstanding Russian mathematician best known for applying differential equations to optimal control, and his popular books on mathematics.

On a personal note, in my early teens, my mentor, the geometer István Reimann, introduced me to the Yaglom–Boltyanskiĭ book, and I absolutely loved it. It may even have helped that, although at the time my Russian was very poor, I had to read this masterpiece in Russian, since that made me prove the results before 'reading' them. To my shame, I needed David Eppstein's reminder that the second proof above was in the Yaglom–Boltyanskiĭ book.

References

Bretschneider, C.A., Untersuchung der trigonometrischen Relationen des geradlinigen Viereckes, *Archiv der Mathematik und Physik* **2** (1842) 225–261.

Hill, Tho., *A Second Book in Geometry*, Brewer and Tileston (1863).

Hobson, E.W., *A Treatise on Plane Trigonometry*, Cambridge University Press (1918).

Yaglom, I.M. and V.G. Boltyanskiĭ, *Vypuklye Figury* (in Russian), State Publishing House (1951); English Translation: *Convex Figures*, Holt, Rinehart and Winston (1961).

24. Constructing $\sqrt[3]{2}$ – Philon of Byzantium

Let OS_1PS_2 be a $2m \times m$ rectangle with $OS_1 = PS_2$ having length $2m$ and $OS_2 = PS_1$ length m. Let C be the circle through the vertices O, S_1, P and S_2, and let Q be the point of the PS_2 arc of C such that the line through P and Q meets the (extended) lines OS_1 and OS_2 in R_1 and R_2, and the segments PR_1 and QR_2 have the same length. Finally, let T_1 and T_2 be the projections of Q on the segments OR_1 and OR_2. Then the segment OT_1 has length $\sqrt[3]{2}m$.

Proof. In addition to $PS_1 = m$ we set $PR_1 = \ell$ and $S_1R_1 = n$, as in Figure 25. Then $QR_2 = \ell$ by definition, so $T_2R_2 = m$ and $T_2Q = n$. Since $OQ \perp QP$, our setup has numerous (eight?) right-angled triangles similar to PS_1R_1, including QT_2O, with the long leg to short leg ratio n/m, and several equalities of segments. In particular:

(i) $OT_1 = T_2Q = n$ and $OS_2 = S_1P = m$, so $OT_1/OS_2 = n/m$;

(ii) $OT_1 = T_2Q$, so $OT_2/OT_1 = OT_2/T_2Q = n/m$;

(iii) $OS_1 = S_2P$ and $OT_2 = S_2R_2$, so $OS_1/OT_2 = S_2P/S_2R = n/m$.

Consequently,

$$(n/m)^3 = \frac{OT_1}{OS_2} \cdot \frac{OT_2}{OT_1} \cdot \frac{OS_1}{OT_2} = \frac{OS_1}{OS_2} = 2,$$

so $OT_1/OS_2 = n/m = \sqrt[3]{2}$ and $OT_1 = \sqrt[3]{2}m$, as claimed. □

Notes. The result above is due to Philon of Byzantium (ca. 280BC – ca. 220BC), an extremely inventive Greek engineer, physicist and mathematician: this was his attempt to solve the ancient and very famous *Delian Problem*, i.e. the problem of 'doubling the cube', i.e. finding the edge of a cube whose volume is twice that of a cube with a given edge. By now we know that there is no ruler-and-compass solution of this problem, but the ancients were absolutely fascinated by it. The line PQ appearing in Philon's construction is *Philon's line*.

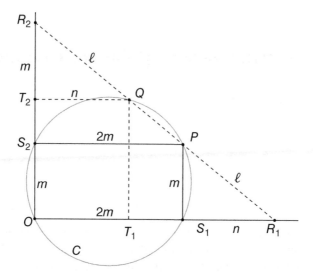

Figure 25 The $2m \times m$ rectangle OS_1PS_2 with its circumcircle C and some additional points.

According to Eratosthenes of Cyrene (ca. 276BC – ca. 195BC), 'when the god proclaimed to the Delians by the oracle that, if they would get rid of the plague, they should construct an altar double the existing one, their craftsmen fell into great perplexity in their efforts to discover how a solid could be made double of a (similar) solid; they therefore went to ask Plato about it, and he replied that the oracle meant, not that the god wanted an altar of double the size, but that he wished, in setting them the task, to shame the Greeks for their neglect of mathematics and their contempt for geometry'.

As Sir Thomas Heath tells us (p. 246), there is no doubt that the question was studied in (Plato's) Academy, and solutions were attributed to Eudoxus, Menaechmus and even (though erroneously) to Plato himself. Later many more people suggested solutions, including Nicodemes, Diocles, Sporus and Pappus. Eratosthenes himself offered a mechanical solution. Philon's 'solution' above may well be the nicest: it constructs $\sqrt[3]{2}$ by taking the intersection Q of the hyperbola $xy = 2$ and the circle C in our figure. (See also Heath (1921), pp. 262–263.) For recent results concerning Philon lines, see the papers of Coxeter and van de Craats, and Wetterling.

An extension of Philon's problem above is the following basic extremal problem. Let ℓ_1 and ℓ_2 be two lines meeting in a point O, and let P be another point not on either line. Determine the line ℓ through P whose segment between the lines ℓ_1 and ℓ_2 has minimal length.

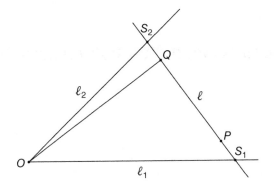

Figure 26 The minimal segment S_1S_2 through P.

A solution is that if ℓ meets ℓ_1 and ℓ_2 in S_1 and S_2 then S_1S_2 is minimal if $PS_1 = QS_2$, where Q is the foot of the perpendicular from O to ℓ, as in Figure 26. Newton offered a different solution: there is a point C such that $CS_1 \perp \ell_1$, $CS_2 \perp \ell_2$ and $CP \perp \ell$.

Fifty years ago, in the 'good old days', when scholarships to Trinity and other Cambridge colleges were awarded on the basis of six 3-hour papers, the only one that really mattered was the geometry paper, with questions like this. Whoever did well in that paper deserved a scholarship. Later I used this question in scholarship interviews: I remember that the brilliant student John Rickard did it easily, but the others found it too hard to do on the spot.

References

Coxeter, H.S.M. and J. van de Craats, Philon lines in non-Euclidean planes, *J. Geometry* **48** (1993) 26–55.

Heath, Sir Thomas, *A History of Greek Mathematics: Vol. I, From Thales to Euclid*, Clarendon Press (1921).

Wetterling, W.W.E., Philon's line generalized: An optimization problem from geometry, *J. Optim. Theory Appl.* **90** (1996) 517–521.

25. Circumscribed Quadrilaterals – Newton

Let ABCD be a quadrilateral circumscribed about a circle with centre O. Let E and F be the midpoints of the diagonals AC and BD. Then E, F and O are collinear.

Proof. If $ABCD$ is a parallelogram, then it is in fact a rhombus, so the points E, F and O coincide. From now on we assume that this is not the case.

As there is a circle inscribed into $ABCD$ (and so touching all four sides), $AB + CD = BC + DA$, so $AOB + COD = BOC + DOA$, where XYZ denotes not only a triangle, but also the area of that triangle.

We shall prove the following assertion, which is more than is required.

Assertion Let $ABCD$ be a convex quadrilateral which is not a parallelogram, and let E and F be the midpoints of the diagonals AC and BD. Then the locus of the points O of $ABCD$ such that $AOB + COD = BOC + DOA$ is the intersection of the line EF with the quadrilateral.

Proof. Note that $AOB + COD = BOC + DOA$ if and only if each of $AOB + COD$ and $BOC + DOA$ is half of $ABCD$, the area of our quadrilateral. This clearly holds if O is replaced by either E or F.

(i) First we show that if O is a point of our quadrilateral on the line ℓ through E and F then $AOB + COD = BOC + DOA$, i.e. $AOB + COD = \frac{1}{2}ABCD$. To prove this, note that A and C are at the same distance from ℓ, and so are B and D. Consequently, $AOE = COE$ and $BOE = DOE$. Assume, as we may, that O is in the triangle ABC, as in Figure 27. Then

$$AOB = AEB + BOE - AOE$$

and, similarly,

$$COD = CED + COE - DOE.$$

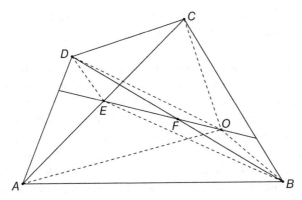

Figure 27 A quadrilateral $ABCD$, with E and F the midpoints of the diagonals AC and BD, and a point O on the line EF.

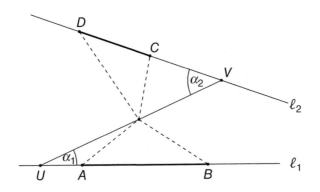

Figure 28 Two segments on two lines that are not parallel, and a segment UV joining them.

Adding these two equations, we find that

$$AOB + COD = AEB + CED + (BOE - DOE) + (COE - AOE)$$
$$= AEB + CED = \frac{1}{2}ABCD,$$

completing the proof of part (i).

(ii) To complete the proof, we have to show that if O is not on ℓ then $AOB + COD \neq \frac{1}{2}ABCD$. Because of (i), this follows from the following obvious assertion.

Let A, B and U be three points on a line ℓ_1 and C, D and V on a line ℓ_2, not parallel to ℓ_1. Let α_i be the angle between ℓ_i and UV, $i = 1, 2$. For a point X of UV, define $f(X) = ABX + CDX$. Then $f(X)$ is strictly monotone unless

$AB(\sin \alpha_1) = CD(\sin \alpha_2)$, in which case it is constant (see Figure 28). Indeed, $2f(X) = AB \cdot UX(\sin \alpha_1) + CD \cdot XV(\sin \alpha_2)$.   □

Notes. This is one of Sir Isaac Newton's many smaller results in geometry. It would be insulting to him and to the reader to give a two-line description of his achievements, so let me just say that he was born on 25th December 1642 according to the Old Style (Julian) calendar, which converts to 4th January 1643 according to the New Style (Gregorian) calendar. For this reason, nowadays his dates are frequently given as 1643–1727 (NS). If in the 16th century the Anglican bishops had not blocked the suggestion of John Dee (the first mathematician at Trinity College, and so one of Sir Isaac Newton's predecessors) to Queen Elizabeth to introduce the Gregorian calendar in Britain, no such confusion would have arisen.

26. Partitions of Integers

(i) *The generating function of the partition function $p(n)$ is*

$$\sum_{n=0}^{\infty} p(n)x^n = \frac{1}{1-x} \cdot \frac{1}{1-x^2} \cdot \frac{1}{1-x^3} \cdots .$$

(ii) *The number of partitions of n that do not contain 1 is $p(n) - p(n-1)$.*

Proof. (i) Consider the following infinite product of formal power series:

$$(1 + x + x^2 + x^3 + \cdots)(1 + x^2 + x^4 + x^6 + \cdots)(1 + x^3 + x^6 + x^9 + \cdots) \cdots .$$

In the expansion of this product, from all but finitely many factors we take the term 1, and from the rest we take x to a non-zero power, and take the sum of all these terms. What is the coefficient of x^n in this expansion? The number of ways x^n can be written as the product of a monomial $x^{m_1 \cdot 1}$, coming from the first factor, and a monomial $x^{m_2 \cdot 2}$, coming from the second factor, and so on, and finally a monomial $x^{m_n \cdot n}$, coming from the nth factor. The condition these multiplicities m_k have to satisfy is $\sum_1^n m_k k = n$. Needless to say, there is no reason to take the monomial 1 from any of the factors ('parentheses'), only the powers of x of degree at least 1. For the same reason, there is no reason to bother about the other (infinitely many) parentheses, since from each we can take only the term 1. This product $x^{m_1 \cdot 1} x^{m_2 \cdot 2} \cdots x^{m_n \cdot n}$ corresponds to the partition of n in which k occurs exactly m_k times. Hence the coefficient of x^n in our expansion is exactly $p(n)$.

Finally, the product of the two formal power series $1 + x^k + x^{2k} + x^{3k} + \cdots$ and $1 - x^k$ is 1, so our proof of part (i) is complete.

(ii) *First Proof* Write $p_1(n)$ for the number of partitions of n not containing 1. A partition is like this if and only if the multiplicity m_1 of 1 in the argument above is 0: hence, $p_1(n)$ is the coefficient of x^n in the expansion of

$$(1 + x^2 + x^4 + x^6 + \cdots)(1 + x^3 + x^6 + x^9 + \cdots) \cdots ,$$

so that

$$\sum_{n=0}^{\infty} p_1(n)x^n = \prod_{k=2}^{\infty} \frac{1}{1-x^k}.$$

Since

$$\prod_{k=1}^{\infty} \frac{1}{1-x^k} = \sum_{k=0}^{\infty} p(n)x^n,$$

we find that

$$\sum_{n=0}^{\infty} p_1(n)x^n = (1-x)\sum_{n=0}^{\infty} p(n)x^n.$$

On the right-hand side the coefficient of x^n is $p(n) - p(n-1)$, so $p_1(n) = p(n) - p(n-1)$, as claimed.

Second Proof Adding 1 to a partition of $n-1$ we get a one-to-one correspondence between the set of all $p(n-1)$ partitions of $n-1$ and the set of partitions of n that contain 1. Hence the number of partitions of n that do not contain 1 is $p(n) - p(n-1)$.

Third Proof This will be a more detailed study of partitions than in the previous proof. To reduce the clutter, in our illustrations of partitions we may omit the $+$ sign, writing 775422111 instead of $7 + 7 + \cdots$.

If a partition of n contains 1 with multiplicity $m \geq 0$ then this partition is the unique concatenation of a partition of $n - m$ that does not contain 1 as a part, and the summand 1 repeated m times, where $0 \leq m \leq n$. (For example, the partition 55322111 of 20 is the concatenation of the partition 55322 of 17 and the trivial partition 111 of 3.) Consequently,

$$p(n) = p_1(n) + p_1(n-1) + \cdots + p_1(2) + p_1(1) + p_1(0)$$
$$= p_1(n) + p_1(n-1) + \cdots + p_1(2) + 1, \tag{1}$$

since $p_1(1) = 0$ and $p_1(0) = 1$. (Indeed, 1 has no partition without 1 as a summand, and the unique partition of 0, the 'empty partition', does not have 1 as a part.)

And now to our task, to show that $p_1(n) = p(n) - p(n-1)$ for $n \geq 2$. We prove this by induction on n. Trivially, $p_1(2) = 1$ and $p(2) - p(1) = 2 - 1 = 1$, so our assertion holds for $n = 2$. Assuming that it holds for $2, \ldots, n-1$, by (1) we have

$$p(n) = p_1(n) + (p(n-1) - p(n-2)) + (p(n-2) - p(n-3))$$
$$+ \cdots + (p(2) - p(1)) + 1 = p_1(n) + p(n-1),$$

completing our third proof of part (ii), and so the solution of our problem is also complete. $\qquad \square$

Notes. The theory of partitions is a vast subject – the two problems above and the few more we shall have do not even scratch the surface of this theory: they are hardly enough to whet the appetite of the reader. The notation used varies greatly: our choice is bound to be somewhat eclectic.

Several of the greatest mathematicians, including Euler, Cayley, Sylvester, Hardy and Ramanujan, paid more than a fleeting tribute to partitions. The mathematician whose interest in the subject immediately put the theory of partitions on the map was Leonhard Euler (1707–1783), one of the greatest mathematicians ever, perhaps *the* greatest.

Multiplicative number theory, dealing with questions of factorization, divisibility, prime numbers, and so on, goes back more than 2000 years to Euclid, but additive number theory, in a proper sense, began with Euler less than 300 years ago: he devoted the sixteenth chapter of his famous treatise, *Introductio in Analysin Infinitorum* (1748), to problems of additive number theory. It was Euler who defined a *partition* as a decomposition of a natural number into summands which themselves are natural numbers, disregarding the order of summation. Exciting questions arise by imposing various conditions on the summands: they may be demanded to be all different, or primes, or cubes, etc.

Euler's theory is based on the fact that algebraical multiplication of integral powers of the same base is the same as the addition of the powers. As we have seen, the number of partitions of n is exactly the coefficient of x^n in the expansion of $1/(1 - x)(1 - x^2) \cdots$ as a power series in x. Euler firmly linked the theory of partitions to the study of various power series. Although we are dealing with *formal* power series, as it happens, these series are absolutely convergent for $|x| < 1$, even if x takes complex values. That these series are convergent is important in other results, but not in the beautiful and simple phenomena we consider.

There are several books on partitions: a significant part of Major MacMahon's treatise is about them, and Andrews has a beautiful introduction.

References

Andrews, G.E., *The Theory of Partitions*, Addison-Wesley Publishing (1976). Reprinted in the series *Cambridge Mathematical Library*, Cambridge University Press (1998).

Euler, L., *Introductio in Analysin Infinitorum* (1748); see also Euler Archive: https://scholarlycommons.pacific.edu/euler-works/102.

MacMahon, Major P.A., *Combinatory Analysis*, volumes I and II, Cambridge University Press (1916). Reprinted with *An Introduction to Combinatory Analysis* (1920), Dover Publications (2004).

27. Parts Divisible by m and $2m$

The number of partitions of n in which no multiple of m is repeated is equal to the number of partitions of n without a multiple of 2m.

Proof. The generating function of partitions without a multiple of $2m$ is

$$\left(\prod_{2m \nmid k}(1 - x^k)^{-1}\right) = \left(\prod_k(1 - x^k)^{-1}\right)\left(\prod_k(1 - x^{2mk})\right),$$

and the generating function of partitions in which no multiple of m is repeated is

$$\left(\prod_{m \nmid k}(1 - x^k)^{-1}\right)\left(\prod_k(1 + x^{km})\right).$$

This latter power series can be rewritten as

$$\left(\prod_k(1 - x^k)^{-1}\right)\left(\prod_k(1 - x^{mk})\right)\left(\prod_k(1 + x^{km})\right)$$

$$= \left(\prod_k(1 - x^k)^{-1}\right)\left(\prod_k(1 - x^{2mk})\right),$$

completing our proof.   □

Note. This is a particularly simple application of encoding partitions by power series.

28. Unequal vs Odd Partitions

(i) *The number of partitions of n into unequal parts is equal to the number of partitions into odd parts.*

(ii) *For m ≥ 1, the number of partitions of n in which no part is repeated more than m times is equal to the number of partitions in which no part is a multiple of m + 1.*

Proof. As suggested in the *Hint*, we prove only part (ii): for a large value of *d* it is harder to miss the easy proof. The generating function of partitions in which no part occurs more than *m* times is

$$\prod_k (1 + x^k + x^{2k} + \cdots + x^{mk}) = \prod_{k=1}^{\infty} \frac{1 - x^{(m+1)k}}{1 - x^k}.$$

We know from Problem 26 that the generating function of all partitions, i.e. the generating function of the partition function $p(n)$, is

$$\prod_{k=1}^{\infty} (1 - x^k)^{-1},$$

with the factor $(1 - x^k)^{-1}$ taking care of the parts equal to k (occurring any number of times). Hence the generating function of the number of partitions in which no part is a multiple of $m + 1$ is this product without the values of k divisible by $m + 1$, i.e.

$$\prod_{k=1}^{\infty} (1 - x^k)^{-1} \prod_{k=1}^{\infty} (1 - x^{(m+1)k}) = \prod_{k=1}^{\infty} \frac{1 - x^{(m+1)k}}{1 - x^k},$$

completing our proof. □

115

Notes. The identity,

$$\prod_{k \geq 1}(1 + x^k) = \prod_{k \geq 1}(1 - x^{2k-1})^{-1},$$

we have proved in the case $m = 1$ is often called *Euler's identity*. Although it is very simple, I still find it surprising.

29. Sparse Bases

There is a set $S \subset \mathbb{N}$ of density zero such that every positive rational is the sum of a finite number of reciprocals of distinct terms of S.

Proof. Let us start by proving the following Claim stated in the *Hint*. It is easily seen that this Claim is essentially equivalent to the assertion we have to prove.

Claim *Let r be a positive rational and A a positive integer. Then there is a finite set $S(r, A) = \{n_1, \ldots, n_\ell\}$ of positive integers such that $n_1 \geq A$, $n_{i+1} - n_i \geq A$ for every i, $i = 1, \ldots, \ell - 1$, and*

$$r = \sum_1^\ell \frac{1}{n_i}.$$

To prove this Claim, first note that there is an integer m such that

$$r - \left(\frac{1}{A} + \frac{1}{2A} + \cdots + \frac{1}{mA} \right) < \frac{1}{(m+1)A}.$$

Writing r_0 for the left-hand side, the Sylvester representation of r_0 (see the end of the Notes in Problem 2, Vulgar Fractions) tells us that there are positive integers $n_1 < \cdots < n_\ell$ such that

$$r_0 = \sum_1^\ell \frac{1}{n_i} \qquad \text{and} \qquad n_{i+1} > n_i(n_i - 1) \quad \text{for every } i, \ 1 \leq i < \ell - 1.$$

Now, since $r_0 < 1/A$, the first term, n_1, is at least $A + 1$, and so

$$n_{i+1} - n_i \geq n_i(n_i - 1) + 1 - n_i = n_i(n_i - 2) + 1 \geq A^2 \geq A,$$

completing the proof of our Claim.

The proof of our assertion is just about immediate: enumerate the positive rationals, and for each rational r, take a finite set $S(r, A)$, with larger and larger values for A. The union of these sets will do for S.

To spell it out, let $r_1, r_2, \ldots$ be an enumeration of the positive rational numbers. Set $A_1 = 1$ and $S_1 = S(r_1, A_1)$. Having defined $S_1, \ldots, S_k$, let s_k be the maximal element of S_k, and set $A_{k+1} = 2s_k$ and $S_{k+1} = S(r_{k+1}, A_{k+1})$. Then the sets $S_1, S_2, \ldots$ are disjoint and $S = \bigcup_i^\infty S_i$ has the required properties: every positive rational number is the sum of the reciprocals of finitely many terms of S, and the density of S is zero, since in the set $\bigcup_{k+1}^\infty S_i$ any two terms are at distance at least 2^k from each other.   □

Notes. A sequence of positive integers $S\{n_1, n_2, \ldots\}$ with $n_1 < n_2 < \cdots$ is an R-basis for the natural numbers if every positive integer is the sum of reciprocals of finitely many integers of S. In 1961, Wilf raised several questions about R-bases. One of these questions was whether an R-basis can have zero density. The result in this problem, proved by Erdős and Stein, goes way beyond answering this question: there is an R-basis for the positive rational numbers (with the obvious definition) that has zero density.

Erdős and Stein write that considerably more is true. For example, the set of reciprocals, $\{1/1, 1/2, 1/3, \ldots\}$, can be partitioned into finite sets such that each positive rational is the sum of the elements of precisely one part. They also prove that there are R-bases $\{n_1, n_2, \ldots\}$ for the positive rationals with $\sum_1^\infty 1/n_i$ tending to infinity as slowly as one likes. Here is the precise statement of this result.

Let $0 < a_1 < a_2 < \cdots$ be such that $\sum_1^\infty 1/a_i = \infty$. Then there is an R-basis $\{n_1, n_2, \ldots\}$ for the positive rational numbers such that $n_i \geq a_i$ for every i.

The proof of this result needs considerably more than the argument we have presented.

References

Erdős, P. and S. Stein, Sums of distinct unit fractions, *Proc. Amer. Math. Soc.* **14** (1963) 126–131.
Wilf, H.S., Reciprocal bases for integers, *Bull. Amer. Math. Soc.* **67** (1961) 456.

30. Small Intersections – Sárközy and Szemerédi

*Let $A_1, \ldots, A_m \in [n]^{(r)}$ be such that $|A_i \cap A_j| \le s < r^2/n$ for all $1 \le i < j \le m$.
Then*

$$m \le \frac{n(r-s)}{r^2 - sn}. \tag{1}$$

*In particular, if r^2/n is an integer, $A_1, \ldots, A_m \in [n]^{(r)}$, and $|A_i \cap A_j| < r^2/n$
for all $1 \le i < j \le m$ then*

$$m \le r - r^2/n + 1 \le n/4 + 1. \tag{2}$$

Proof. As a preliminary observation, note that if A and B are random r-subsets
of $[n]$, then the expected size of their intersection is r^2/n. Thus our condition
says that all pairwise intersections are smaller, perhaps significantly smaller,
than in the random case. This is the reason why one cannot choose all that many
sets A_i.

And now let us turn to our proof. For $1 \le i \le m$, set $f_i = \mathbf{1}_{A_i}$, i.e. let
$f_i : [n] \to \mathbb{R}$ be given by

$$f_i(h) = \begin{cases} 1 & \text{if } h \in A_i, \\ 0 & \text{otherwise,} \end{cases}$$

and set $F = \sum_{i=1}^m f_i$.

The conditions on the sets A_i translate into

$$\sum_{h=1}^n f_i(h) = r$$

119

for every i, $1 \leq i \leq n$, and

$$\sum_{h=1}^{n} f_i(h)f_j(h) \leq s < r^2/n$$

whenever $i \neq j$. In particular,

$$\sum_{h=1}^{n} F(h) = \sum_{h=1}^{n} \sum_{i=1}^{m} f_i(h) = \sum_{i=1}^{m} \sum_{h=1}^{n} f_i(h) = rm.$$

Hence, by the Cauchy–Schwarz inequality,

$$rm = \sum_{h=1}^{n} F(h) \leq n^{1/2} \left(\sum_{h=1}^{n} F(h)^2 \right)^{1/2},$$

so

$$\sum_{h=1}^{n} F(h)^2 \geq r^2 m^2/n. \tag{3}$$

On the other hand, we can bound $\sum_{h=1}^{n} F(h)^2$ from above as follows:

$$\sum_{h=1}^{n} F(h)^2 = \sum_{h=1}^{n} \left(\sum_{i=1}^{m} f_i(h) \right) \left(\sum_{j=1}^{m} f_j(h) \right) = \sum_{i \neq j} \sum_{h=1}^{n} f_i(h)f_j(h) + \sum_{h=1}^{n} \sum_{i=1}^{m} f_i(h)^2$$

$$\leq m(m-1)s + rm.$$

Hence, recalling inequality (3),

$$r^2 m^2/n \leq m(m-1)s + rm,$$

i.e.

$$m(r^2/n - s) \leq r - s,$$

proving inequality (1).

To see the rider, it suffices to note that if r^2/n is an integer and $|A_i \cap A_j| < r^2/n$ then $|A_i \cap A_j| \leq r^2/n - 1$, so the conditions hold with $s = r^2/n - 1$: this implies the first inequality in (2). The second follows since for $r \geq 1$, the expression $r - r^2/n$ attains its maximum at $r = n/2$. □

Notes. This is one of the results Sárközy and Szemerédi proved in 1970. There are a host of results on this topic: for a sample, see the paper by Sárközy and Sárközy.

References

Sárközy, A. and G.N. Sárközy, On the size of partial block designs with large blocks, *Discrete Math.* **305** (2005) 264–275.

Sárközy, A. and E. Szemerédi, On intersections of subsets of finite sets (in Hungarian), *Mat. Lapok* **21** (1970) 269–278.

31. The Diagonals of Zero–One Matrices

The maximal number of diagonals of zero–one matrices that one can obtain by permuting the rows of an $n \times n$ matrix is $2^n - n$.

Proof. First, let us show that at least n different zero–one sequences *will not appear* as diagonals of the matrices obtained by permuting the rows of an $n \times n$ matrix $A = (a_{ij})_{i,j=1}^{n}$ with $a_{ij} = 0, 1$. To this end, note that for each i, $1 \le i \le n$, the diagonal of a matrix obtained from A by permuting its rows agrees with each row in at least one place. (Here a row is considered to be a sequence of length n and so is the diagonal.)

Hence, if no two rows are the same, then at least n diagonals cannot be obtained. On the other hand, if some two rows are the same, say, $R_1 = R_2 = (a_1, a_2, \ldots, a_n)$, then the $n+1$ sequences agreeing with $(1-a_1, 1-a_2, \ldots, 1-a_n)$ in all but at most one place fail to appear as diagonals in the permutations.

It is even easier to see that there need not be more than n zero–one sequences that will *not* be the diagonals of our permuted matrix: simply let A be the identity matrix with 1s on the diagonal and 0s elsewhere. Then for every k, $2 \le k \le n$, and all sequences $1 \le i_1 < i_2 < \cdots < i_k \le n$, permuting the rows with these indices cyclically (so that the i_1th row becomes the i_2th, the i_2th becomes the i_3th, etc., finally the i_kth row the i_1th) the diagonal of the new matrix has a 0 in precisely the places indexed by $i_1, i_2, \ldots, i_{k-1}$ and i_k. $\qquad \square$

32. Tromino and Tetronimo Tilings

(i) *If n is a power of 2 then every deficient $n \times n$ board can be tiled with trominoes.*

(ii) *If an $m \times n$ rectangle can be tiled with T-tetrominoes then mn is divisible by 8.*

Proof. (i) Let us apply induction on n, starting with the trivial case $n = 1$, when no trominoes are needed for our tiling. Assuming that $n = 2^{k+1}$ and the assertion holds for 2^k, divide the $n \times n$ grid into four $2^k \times 2^k$ grids. The cell we have removed has been removed from exactly one of these quarters. Put a tromino in the 'centre' of the original grid that meets each of the other three (full) quarters. All that remains is to tile each of the quarters with one cell removed.

(ii) Since each T-tetromino consists of four cells, mn is divisible by 4. Hence, we may assume that m is even. To go further, consider a black-and-white checkerboard colouring of our $m \times n$ rectangle tiled with T-tetrominoes. Since each row has an even number of squares, half of the squares are white and half are black; consequently, the same applies to the entire board.

Also, in the T-tetrominoes in our tiling, each T-tetromino has either three black cells and one white cell, or three white cells and one black cell; consequently, our tiling must contain the same number, $mn/8$, of each kind. In particular, 8 divides mn. $\qquad\square$

Notes. A *polyomino of order k* is the union of k cells (unit squares) of a grid joined at the edges. Thus a tromino is a polyomino of order 3, and a T-tetromino is a polyomino of order 4. Polyominoes were introduced by Golomb in 1954, and since then they have been much studied in recreational mathematics. The first lovely (though trivial) assertion in this problem above is also due to Golomb. In fact, much more is true. First, the proof above shows that the assertion holds in all dimensions. Thus, if n is a power of 2 then the d-dimensional $n \times n \times \cdots \times n$ board can be tiled with deficient d-dimensional $2 \times 2 \times \cdots \times 2$ polyominoes. Also, Chu and Johnsonbaugh have proved the more difficult result that a deficient $n \times n$ board can be tiled with trominoes if and only if n is not a multiple of 3 (of course!) and $n \neq 5$.

Deciding for which pairs (m, n) a (full) $m \times n$ rectangle can be tiled with trominoes is even easier. Trivially, mn must be a multiple of 3, and if $m = 3$, say, then n cannot be odd. In every other case, there is a tiling with trominoes. To see that, it helps to use (the obvious) tilings of 3×5 and 5×9 rectangles as building blocks.

Part (ii) is also true in a *much* stronger form. Walkup proved in 1965 that an $m \times n$ rectangle can be tiled with tetrominoes if and only if both m and n are divisible by 4. The sufficiency of these conditions is obvious, since a 4×4 square has a tiling, but the necessity needs work.

Closer to home, the second problem is an elder brother of the classical puzzle that an $n \times n$ square with two diagonally opposite unit squares removed cannot be tiled with dominoes. As just about everyone knows, the standard solution is based on parity: we take a chessboard colouring of the board, and note that we have removed two squares of the same colour. I first heard of the T-tetromino question above in my early teens from the Hungarian geometer Dr István Reiman, who was my first mentor in mathematics. Later he became the Leader of the IMO (International Mathematical Olympiad) Team of Hungary. When in the 1970s I interviewed scholarship candidates for Trinity College, Cambridge, after the introductory question about domino tilings, I asked them the tetromino question above, with varying degrees of success.

To end these notes, we turn to more substantial aspects of T-tetromino tilings. In 2004, Korn and Pak proved that any two T-tetromino tilings of a rectangular board are connected by moves involving two or four tiles. Furthermore, the number of such tilings is an evaluation of the Tutte polynomial, a very important polynomial in graph theory and statistical physics.

References

Chu, I.P. and R. Johnsonbaugh, Tiling deficient boards with trominos, *Math. Mag.* **59** (1986) 34–40.

Golomb, S.W., Checker boards and polyominoes, *Amer. Math. Monthly* **61** (1954) 675–682.

Golomb, S.W., *Polyominoes – Puzzles, Patterns, Problems, and Packings*, with diagrams by Warren Lushbaugh, Second edition; with an appendix by Andy Liu, Princeton University Press (2020).

Korn, M. and I. Pak, Tilings of rectangles with T-tetrominoes, *Theoret. Comput. Sci.* **319** (2004) 3–27.

Walkup, D.W., Covering a rectangle with T-tetrominoes, *Amer. Math. Monthly* **72** (1965) 986–988.

33. Tromino Tilings of Rectangles

An $m \times n$ rectangle can be tiled with trominoes if and only if $m, n \geq 2$ and mn is a multiple of 3, except when one is 3 and the other is odd.

Proof. To reduce the number of cases, we shall often make use of the fact that m and n are interchangeable.

(i) *The necessity of the conditions.* Since a tromino consists of three cells (unit squares) glued together, $m, n \geq 2$ and 3 divides mn. Suppose $m = 3$ and we have a tiling of the $3 \times n$ rectangle (with trominoes). Consider the tiles touching the left side (of length 3) of our $3 \times n$ rectangle, as in Figure 29. There must be two tiles which, together, tile the 3×2 rectangle on the left of the $3 \times n$ rectangle, so the remaining trominoes tile the $3 \times (n - 2)$ rectangle on the right of our $3 \times n$ rectangle. By induction on n, this shows that n is even.

(ii) *The sufficiency of the conditions.* As shown in the figure where the problem is posed, the 3×2 and 5×9 rectangles can be tiled, and so can the 2×3 and 9×5 rectangles. Indeed, to tile the 5×9 rectangle, first place five 2×3 and 3×2 rectangles into a 5×9 rectangle, as in Figure 30, and then tile the rest with five tiles.

Suppose that 3 divides m, so that $m = 3\ell$, and if $m = 3$ then n is even. A $3 \times (2k)$

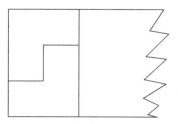

Figure 29 The two tiles touching the left side of a $3 \times n$ rectangle.

126

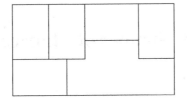

Figure 30 Placing 2 × 3 and 3 × 2 rectangles into a 5 × 9 rectangle.

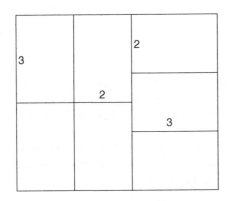

Figure 31 A tiling of a 6 × 7 rectangle with 2 × 3 and 3 × 2 rectangles.

rectangle can be partitioned into 3 × 2 rectangles (*k* of them), so can be tiled. Hence, if *n* is even, our *m* × *n* rectangle can be tiled with trominoes. From now on, suppose that $n \geq 3$ is odd and $m = 3\ell \geq 6$. Furthermore, if $n = 3$ then *m* is even. Our task then is to show that an *m* × *n* rectangle can be tiled (with trominoes).

A 6 × *n* rectangle can be partitioned into 3 × 2 and 2 × 3 rectangles, as in Figure 31, so can be tiled with trominoes.

Furthermore, since the 9 × 2 and 9 × 5 rectangles can be tiled with trominoes, so can every 9 × *n* rectangle, with $n \geq 5$ odd. Hence, if $k \geq 2$ and *n* is odd, every 3*k* × *n* rectangle can be tiled. Indeed, write *k* as

$$k = 2r + 3s, \qquad r, s \geq 0; \ r \neq 1,$$

and partition our 3*k* × *n* rectangle into *r* rectangles, each 6 × *n*, and *s* rectangles, each 9 × *n*. As we have seen, each of these rectangles can be tiled with trominoes, completing our proof. □

34. Number of Matrices

Let $\mathcal{A}_{3,7}^{(n)}$ be the set of $n \times n$ matrices with non-negative integer entries, in which every row and column has at most three non-zero entries, such that non-zero entries are different, and their sum is 7. Then $\mathcal{A}_{3,7}^{(n)}$ consists of $(n!)^3$ matrices.

Proof. The non-zero entries of a row or column of a matrix $A \in \mathcal{A}_{3,7}^{(n)}$ can be $(4,2,1)$, $(5,2)$, $(6,1)$ and (7). The latter three can be written *uniquely* as sums of 1, 2 and 4, so that each grouping contains all three numbers:

$$(5,2) = (1+4,2), \qquad (6,1) = (2+4,1) \qquad \text{and} \quad (7) = (1+2+4).$$

This shows that for every matrix in $\mathcal{A}_{3,7}^{(n)}$ there is a unique way of writing it as a sum of three permutation matrices with 4, 2 and 1 in place of the usual 1 entry, as in the following example:

$$
\begin{bmatrix} 2 & 4 & 0 & 1 \\ 0 & 0 & 1 & 6 \\ 4 & 3 & 0 & 0 \\ 1 & 0 & 6 & 0 \end{bmatrix}
=
\begin{bmatrix} 0 & 4 & 0 & 0 \\ 0 & 0 & 0 & 4 \\ 4 & 0 & 0 & 0 \\ 0 & 0 & 4 & 0 \end{bmatrix}
+
\begin{bmatrix} 2 & 0 & 0 & 0 \\ 0 & 0 & 0 & 2 \\ 0 & 2 & 0 & 0 \\ 0 & 0 & 2 & 0 \end{bmatrix}
+
\begin{bmatrix} 0 & 0 & 0 & 1 \\ 0 & 0 & 1 & 0 \\ 0 & 1 & 0 & 0 \\ 1 & 0 & 0 & 0 \end{bmatrix}
$$

A matrix as a sum of three multiples of permutation matrices.

Conversely, the sum of three permutation matrices with coefficients 4, 2 and 1 belongs to $\mathcal{A}_{3,7}^{(n)}$, so our proof is complete. $\qquad\qquad \square$

35. Halving Circles

Let S be a set of $2n + 1 \geq 5$ points in the plane in general position. Then S has at least $n(2n + 1)/3$ halving circles.

Proof. Let us show that every pair of points is on a halving circle. As there are $\binom{2n+1}{2} = n(2n + 1)$ pairs of points, and each halving circle contains three pairs, this implies our assertion.

Let p and q be two points of S. As we have $2n - 1$ points other than p and q, one of the two half-planes determined by the line pq contains at least n points of S, and the other at most $n - 1$. Let a and b be the centres of two large circles through p and q that contain the points of S in these half-planes, without a point of S in both circles, as in Figure 32. Note that the segment $[a, b]$ is a perpendicular bisector of $[p, q]$. For a point $c \in [a, b]$, let C_c be the circle with centre c through p and q, and write $n(c)$ for the number of points inside C_c. Since $n(a) + n(b) = 2n - 1$, we may assume that $n(a) \geq n$ and so $n(b) \leq n - 1$.

Moving c continuously from a to b along $[a, b]$, as we start with $n(a) \geq n$ points and end with $n(b) \leq n - 1$ points inside the circle, for some point c_0 the

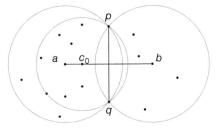

Figure 32 $n = 7, 15$ points. C_{c_0} has three points of S on it, including p and q, and $6 = 7 - 1$ in its interior.

129

circle C_{c_0} will have another point of S on its boundary and precisely $n-1$ points inside. This circle C_{c_0} is thus a halving circle through p and q, as claimed. ☐

Notes. There is no doubt that in this form this problem is on the easy side for this volume: as a folklore result it has been around for ever. We shall return to it in a slightly more sophisticated form in the next problem.

36. The Number of Halving Circles

For every $n \geq 1$ there is a set S of $2n + 1$ points in general position in the plane with exactly n^2 halving circles.

Proof. We shall define S as the union of two fairly similar sets, S_1 and S_2, with $|S_1| = n + 1$ and $|S_2| = n$. To construct them, let $f : \mathbb{R} \to \mathbb{R}$ be a strictly concave function with $f(0) = f(n) = 0$ and $\max_{0 \leq x \leq n} f(x) = f(n/2) = 2^{-n}$. Thus f is close to being identically zero between 0 and n; although this is much more than we need, it costs us nothing. Also, let

$$S_1 = \{(i, y_i) : 0 \leq i \leq n\},$$

where $y_i = f(i)$. We may assume that f has been chosen to guarantee that no four points of S_1 are on a circle. Similarly, let

$$S_2 = \{(i, 2^n - y_i) : 1 \leq i \leq n\}.$$

Thus S_1 and S_2 are on two arcs 'facing' each other, as in Figure 33.

What is the number of halving circles of $S = S_1 \cup S_2$? Each halving circle has two points in S_1 and one in S_2, or two in S_2 and one in S_1. Clearly, for $0 \leq i < j \leq n$, a halving circle through (i, y_i) and (j, y_j) contains precisely $j - i - 1$ points of S_1, and so $n - j + i$ points of S_2; hence, precisely one halving circle passes through any two points of S_1. Analogously, precisely one halving circle passes through any two points of S_2. Therefore, the total number of halving circles of S is

$$\binom{n+1}{2} + \binom{n}{2} = n^2,$$

as required. $\qquad\qquad\qquad\qquad\qquad\qquad\qquad\qquad\qquad\qquad\qquad$ □

Notes. The assertion above should be viewed as part of the much more interesting and surprising result that *any* set of $2n + 1$ points in general position in

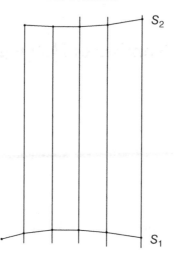

Figure 33 $n = 5$, 11 points, $S = S_1 \cup S_2$; every halving circle has two points in $S - 1$ or two points in S_2.

the plane has exactly n^2 halving circles. This lovely fact was noticed and proved by Ardila in 2004. The natural proof consists of two parts. The first is that any two appropriate sets have the same number of halving circles, and the second that *some* set of $2n + 1 \geq 3$ points in general position in the plane has n^2 halving circles. Thus the present problem is this second part.

In fact, the first part is also easily proved: moving *one* of the points (slowly and in a sensible way), when this moving point crosses only one of the lines and circles determined by our set of points then the number of halving circles remains constant. This is easily seen, only it is slightly fiddly.

The proof of the second part above is different (and somewhat simpler) than the one given by Ardila.

In conclusion, let us note that Ardila proved a more general result as well: he determined the number of circles splitting a $(2n + 1)$-set in any proportion. Given non-negative integers $a \neq b$ with $a + b = 2n - 2$, call a circle through three points of our set S of $2n + 1$ points in general position (a, b)-*splitting* if it has either a or b points in its interior (and so b or a points outside). Then there are precisely $2(a + 1)(b + 1)$ circles that are (a, b)-splitting. For $a = b = n - 1$ this tells us that there are n^2 halving circles, as we have said. Surprisingly, not even the special case $a = 0$ (and $b = 2n - 2$) is trivial, telling us that there are $4n - 2$ circles through three points that contain no point or contain all other points. Note that there can be many circles through three points that contain all other points, although one may be inclined to think that there can be only one.

Reference

Ardila, F., The number of halving circles, *Amer. Math. Monthly* **111** (2004) 586–591.

37. A Basic Identity of Binomial Coefficients

Let $f(X)$ be a polynomial of degree less than n. Then

$$\sum_{k=0}^{n} (-1)^k \binom{n}{k} f(k) = 0.$$

Proof. Trivially,

$$\sum_{k=0}^{n} (-1)^k \binom{n}{k} = (1-1)^n = 0.$$

Consequently, writing, as usual, $(k)_r$ or $k_{(r)}$ for the falling factorial $k(k-1)(k-2)\cdots(k-r+1)$, for $1 \leq r \leq n-1$ we have

$$\sum_{k=0}^{n} (-1)^k \binom{n}{k} k_{(r)} = n_{(r)} \sum_{k=r}^{n} (-1)^k \binom{n-r}{k-r} = 0.$$

Also, every polynomial of degree at most $n-1$ is a linear combination of the falling factorials $X_{(0)} = 1, X_{(1)} = X, X_{(2)}, \ldots, X_{(n-1)}$:

$$f(X) = \sum_{r=0}^{n-1} c_r X_{(r)}.$$

Therefore

$$\sum_{k=0}^{n} (-1)^k \binom{n}{k} f(k) = \sum_{k=0}^{n} (-1)^k \binom{n}{k} \sum_{r=0}^{n-1} c_r k_{(r)}$$

$$= \sum_{r=0}^{n-1} c_r \sum_{k=0}^{n} (-1)^k \binom{n}{k} k_{(r)} = 0,$$

as claimed. □

Notes. There are many identities involving binomial coefficients: for two collections, see the references below. There are several reasons why this particular basic identity has been set as an exercise: its form is very pleasing, its proof is elegant and simple, and the result itself will be needed in Problem 40.

References

Gould, H.W., *Combinatorial Identities*, published by the author (1972).
Riordan, J., *Combinatorial Identities*, reprint of the 1968 original, Robert E. Krieger Publishing Co. (1979).

38. Tepper's Identity

For a natural number n and a real number x, set

$$F_n(x) = \sum_{i=0}^{n} (-1)^i \binom{n}{i} (x-i)^n.$$

Then $F_n(x) = n!$.

Proof. Stating the problem in this form, the secret is out, and there is an obvious line of attack: first, show that the polynomial $F_n(x)$ is a constant function and, second, pick a value of x for which it is reasonably easy to prove that $F_n(x) = n!$. The first is the less easy step: to emphasize it, we state it as a Claim.

Claim *For each natural number n, there is a constant C_n such that $F_n(x) = C_n$.*

We prove this by induction on n, starting with $F_1(x) = x - (x-1) = 1$. For the induction step, it suffices to show that $F_n'(x) = 0$ on an open interval, say. Trivially,

$$F_n'(x) = n \sum_{i=0}^{n} (-1)^i \binom{n}{i} (x-i)^{n-1}$$

$$= n \sum_{i=0}^{n} (-1)^i \left(\binom{n-1}{i} + \binom{n-1}{i-1} \right) (x-i)^{n-1}$$

$$= n \sum_{i=0}^{n-1} (-1)^i \binom{n-1}{i} (x-i)^{n-1}$$

$$- n \sum_{i=1}^{n} (-1)^{i-1} \binom{n-1}{i-1} (x-1-(i-1))^{n-1}$$

$$= nC_{n-1} - nC_{n-1} = 0,$$

proving the Claim.

The second step is even easier. We choose $x = n$, and note that $(n)_{(i)}(n-i) = n(n-1)\cdots(n-1-i)(n-i) = n(n-1)_{(i)}$. Hence,

$$F_n(n) = \sum_{i=0}^{n} (-1)^i \binom{n}{i}(n-i)^n$$

$$= \sum_{i=0}^{n} (-1)^i n \binom{n-1}{i}(n-i)^{n-1} = nC_{n-1},$$

so $F_n(x) = n!$ for every x, completing the proof of Tepper's Identity. □

Notes. This identity was published in 1965 by Myron Tepper as a conjecture, although *immediately* after his note in *Mathematics Magazine*, Calvin Long published a proof of it. When in the early 1970s I set this problem to first-year undergraduates at Trinity, I had in mind the proof above. As it is far from obvious that the polynomial in question is constant, the problem was not a pushover. Today this question could not be set, since every undergraduate would use a laptop, and would find at once that the function is exactly $n!$. And from then on very few of them would have any difficulty.

A proof similar to the one above was given in 1972 by Papp.

References

Long, C.T., Proof of Tepper's factorial conjecture, *Math. Mag.* **38** (1965) 304–305.
Papp, F.J., Another proof of Tepper's identity, *Math. Mag.* **45** (1972) 119–121.

39. Dixon's Identity – Take One

Let a, b, c be non-negative integers. Then

$$\sum_k \frac{(-1)^k (a+b)!(b+c)!(c+a)!}{(a+k)!(a-k)!(b+k)!(b-k)!(c+k)!(c-k)!} = \frac{(a+b+c)!}{a!b!c!},$$

with the convention that $0! = 1$ *and* $1/k! = 0$ *for* $k < 0$.

Proof. Let us apply induction on a, starting with the trivial case $a = 0$, when there is only one term on the left-hand side. Write $F(a, k)$ for the kth term on the left-hand side:

$$F(a,k) = \frac{(-1)^k (a+b)!(b+c)!(c+a)!}{(a+k)!(a-k)!(b+k)!(b-k)!(c+k)!(c-k)!},$$

and $S(a)$ for the left-hand side itself, i.e. the sum $\sum_k F(a, k)$. The induction step follows if

$$(a+1)S(a+1) = (a+b+c+1)S(a), \tag{1}$$

i.e. the difference between the two sides is 0. To prove (1), we shall compare the kth terms in the sums giving $S(a+1)$ and $S(a)$. Define a function $G(a, k)$ as follows:

$$G(a,k) = \frac{(-1)^k (a+b)!(b+c)!(c+a)!}{2(a+1+k)!(a-k)!(b+k)!(b-1-k)!(c+k)!(c-1-k)!}.$$

We claim that

$$(a+1)F(a+1,k) - (a+b+c+1)F(a,k) = G(a,k) - G(a,k-1). \tag{2}$$

Summing the two sides of (2) over k, the telescoping series we get on the right-hand side implies identity (1).

138

Multiplying (2) by

$$\frac{(-1)^k(a+1+k)!(a+1-k)!(b+k)!(b-k)!(c+k)!(c-k)!}{(a+b)!(b+c)!(c+a)!},$$

we see that our task is to show that

$$(a+1) \ (a+b+1)(a+c+1) - (a+1+k)(a+1-k)(a+b+c+1)$$
$$= \frac{1}{2}\left[(a+1-k)(b-k)(c-k) + (a+1+k)(b+k)(c+k)\right].$$

Expanding these expressions, we find that both sides are

$$(a+1)bc + k^2(a+1+b+c),$$

completing the proof of (2), and so of the induction step. □

Notes. Dixon published his inequality in 1903, with a not very attractive proof. In the last fifty years several much simpler proofs have been published: this very short proof is due to Doron Zeilberger, writing as Shalosh B. Ekhad. This proof is not only shorter than the previous proofs, but also assumes no knowledge of mathematics beyond high-school algebra. To emphasize this, we have not used binomial coefficients to express the terms on the left-hand side of the identity. In the next problem we shall give another proof: in that proof the binomial notation will be more natural.

References

Dixon, A.C., Summation of a certain series, *Proc. London Math. Soc.* **35** (1903) 285–289.

Ekhad, S.B., A very short proof of Dixon's theorem, *J. Combin. Theory, Ser. A* **54** (1990) 141–142.

40. Dixon's Identity – Take Two

(i) Let m and n be non-negative integers, and write X for a variable. Then

$$\sum_{k=0}^{2n}(-1)^k\binom{m+2n}{m+k}\binom{X}{k}\binom{X+m}{m+2n-k} = (-1)^n\binom{X}{n}\binom{X+m+n}{m+n}.$$

(ii) Deduce from (i) that if a, b and c are non-negative integers and, say, $b \le a, c$, then

$$\sum_{k=-b}^{b}(-1)^k\binom{a+b}{a+k}\binom{b+c}{b+k}\binom{c+a}{c+k} = \binom{a+b+c}{a,b,c} = \frac{(a+b+c)!}{a!b!c!}.$$

Proof. Write $P(X)$ for the left-hand side of our identity to be proved, and $Q(X)$ for its right-hand side. Then P is a polynomial of degree at most $m + 2n$ and Q is a polynomial of degree $m + 2n$. Consequently, they are identical if they agree at $m + 2n + 1$ places, namely at $-m - n, -m - n + 1, \ldots, n$.

Clearly, $Q(n) = (-1)^n\binom{m+2n}{m+n}$ and $Q(x) = 0$ at the $m + 2n$ places $x = -m - n, -m - n + 1, \ldots, n - 1$. Hence, to prove our polynomial identity, it suffices to show that P is also zero at these $m + 2n$ places, and $P(n) = (-1)^n\binom{m+2n}{m+n}$. The second relation is easily proved: in the expansion of $P(n)$ there is only one non-zero summand, that with $k = n$, so $P(n) = (-1)^n\binom{m+2n}{m+n}$.

To prove that P is zero at each integer value from $-m-n$ to $n-1$, we partition this range into three parts. Let us keep in mind that $0 \le k \le 2n$.

(1) The range $x = 0, 1, \ldots, n - 1$. In this case, if $x \le k - 1$ then $\binom{x}{k}$, the second binomial factor, is zero; otherwise $k \le x \le n-1 < n \le 2n-k$, so $\binom{x+m}{m+2n-k}$, the third factor is zero.

(2) The range $x = -m, -m + 1, \ldots, -1$. In this case $x + m \le m - 1 < m \le m + 2n - k$, so $\binom{x+m}{m+2n-k} = 0$, as above.

140

(3) The range $x = -m - n, -m - n + 1, \ldots, -m - 1$. This case is the heart of the proof. Let us make the substitution $x = -y - 1$, so that $y = m, m + 1, \ldots, m + n - 1$. Then, at these *positive integer* values of y, both $-y - 1$ and $-y - 1 + m$ are negative, so

$$P(-y - 1) = \sum_{k=0}^{2n} (-1)^k \binom{m + 2n}{m + k} \binom{-y - 1}{k} \binom{-y - 1 + m}{m + 2n - k}$$

$$= \sum_{k=0}^{2n} (-1)^{k + k + m + 2n - k} \binom{m + 2n}{m + k} \binom{y + k}{k} \binom{y + 2n - k}{m + 2n - k}$$

$$= \sum_{k=0}^{2n} (-1)^{m+k} \binom{m + 2n}{m + k} \binom{y + k}{k} \binom{y + 2n - k}{y - m}.$$

The product of the second and third binomial coefficients above is a polynomial $f_{m,n,y}(k)$ of k:

$$f_{m,n,y}(k) = \binom{y + k}{k} \binom{y + 2n - k}{y - m}.$$

In our range, i.e. for $y = m, m + 1, \ldots, m + n - 1$, the degree of $f_{m,n,y}(k)$ is $2y - m < m + 2n$. Also, $\binom{y+k}{y} = 0$ if k is negative, so in the formula for $P(-y-1)$ the summation can be extended to negative values of k. Recalling Problem 37, we have

$$P(-y - 1) = \sum_{k=-m}^{2n} (-1)^{m+k} \binom{m + 2n}{m + k} f_{m,n,y}(k) = 0.$$

This completes the proof of part (i).

(ii) Set $m = a - b$ and $n = b$. Then the evaluation of the polynomial identity in part (i) at $X = b + c$ tells us that

$$\sum_{k=0}^{2b} (-1)^k \binom{a + b}{a - b + k} \binom{b + c}{k} \binom{c + a}{a + b - k} = (-1)^b \binom{b + c}{b} \binom{b + c + a}{a}.$$

Substituting k for $k - b$, this gives

$$\sum_{k=-b}^{b} (-1)^k \binom{a + b}{a + k} \binom{b + c}{b + k} \binom{c + a}{c + k} = \binom{a + b + c}{a, b, c},$$

as claimed. $\square$

Notes. The second part of this result is Dixon's Identity, which first appeared in Problem 39. with a short but clever proof. This extension of it to polynomials and the beautiful proof above were given by Guo in 2003.

I consider this to be the conceptually simplest proof to date: after the hint that the two sides of the polynomial identity should be evaluated at the integer values from $-m - n$ to n, at every place other than n the vaue is zero, it should be easy to find a proof.

Although Dixon's Identity has many other proofs in addition to the two simple and beautiful proofs in this book, I doubt that this is the end of the road. As the right-hand side of the original identity (not its extension to polynomials) is the number of ways of partitioning a set with $a + b + c$ elements into three sets A, B and C with $|A| = a$, $|B| = b$ and $|C| = c$, I expect that there is a simple proof by counting in which the inclusion–exclusion formula is used.

References

Dixon, A.C., Summation of a certain series, *Proc. London Math. Soc.* **35** (1903) 285–289.

Ekhad, S.B., A very short proof of Dixon's theorem, *J. Combin. Theory, Ser. A* **54** (1990) 141–142; 388–389.

Gessel, I. and D. Stanton, Short proofs of Saalschütz's and Dixon's theorems, *J. Combin. Theory, Ser. A* **38** (1985) 87–90.

Guo, V.J.W., A simple proof of Dixon's identity, *Discrete Math.* **268** (2003) 309–310.

41. An Unusual Inequality

Let $x_0 = 0 < x_1 < x_2 < \cdots$. Then

$$\sum_{n=1}^{\infty} \frac{x_n - x_{n-1}}{x_n^2 + 1} < \frac{\pi}{2}.$$

Proof. As in Figure 34, let $O = (0,0)$, $X_n = (1, x_n)$, $n \geq 0$, $Y_\infty = (0, 1)$, and denote by C the unit circle with centre O. For $n \geq 1$, let Y_n be the intersection of the segment OX_n with the circle C, and Z_{n-1} the intersection of the line through Y_n parallel with $X_0 X_n$ with the segment OX_{n-1}. For $n \geq 1$, denote by T_n the triangle $OY_n Z_{n-1}$, and write $|T_n|$ for its area. The triangles $OY_n Z_{n-1}$ and $OX_n X_{n-1}$ are homothetic with ratio $|OY_n|/|OX_n| = 1/|OX_n| = 1/\sqrt{x_n^2 + 1}$, and the area of the triangle $OX_n X_{n-1}$ is $|X_n X_{n-1}|/2 = (x_n - x_{n-1})/2$, so

$$|T_n| = \frac{x_n - x_{n-1}}{2(x_n^2 + 1)}.$$

Clearly, the triangles $T_1, T_2, \ldots$ are disjoint (in the sense that no two share an

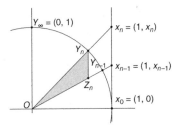

Figure 34 The shaded triangle, T_n, has area $|T_n| = (x_n - x_{n-1})/2(x_n^2 + 1)$.

143

internal point) and are contained in the quarter circle OX_0Y_∞ of area $\pi/4$, so

$$\sum_{n=1}^{\infty} \frac{x_n - x_{n-1}}{2(x_n^2 + 1)} < \pi/4,$$

completing our proof.   □

Notes. An equivalent form of this inequality is the following: let $r > 0$ and $a_0 = 0 < a_1 < \cdots$; then

$$\sum_{n=1}^{\infty} \frac{r(a_n - a_{n-1})}{a_n^2 + r^2} < \frac{\pi}{2}.$$

Indeed, setting $x_n = a_n/r$ in the inequality we have just proved gives

$$\sum_{n=1}^{\infty} \frac{a_n/r - a_{n-1}/r}{(a_n/r)^2 + 1} = \sum_{n=1}^{\infty} \frac{r(a_n - a_{n-1})}{a_n^2 + r^2} < \frac{\pi}{2}.$$

Also, since

$$\sqrt{n} - \sqrt{n-1} = \frac{(\sqrt{n} - \sqrt{n-1})(\sqrt{n} + \sqrt{n-1})}{\sqrt{n} + \sqrt{n-1}} = \frac{1}{\sqrt{n} + \sqrt{n-1}} > \frac{1}{2\sqrt{n}},$$

with $r = \sqrt{m}$ and $a_n = \sqrt{n}$ we find that

$$\sum_{n=1}^{\infty} \frac{\sqrt{m}}{\sqrt{n}(m+n)} < \pi.$$

This inequality will be used in Problem 42, Hilbert's Inequality.

42. Hilbert's Inequality

Let $(a_n)_1^\infty$ and $(b_n)_1^\infty$ be square-summable sequences of real numbers: $\sum_n a_n^2 < \infty$ and $\sum_n b_n^2 < \infty$. Then

$$\sum_{m,n} \frac{a_m b_n}{m+n} < \pi \sqrt{\sum_m a_m^2} \sqrt{\sum_n b_n^2}.$$

Proof. Recalling the final inequality in the solution of Problem 41,

$$\sum_{m=1}^\infty \sum_{n=1}^\infty \frac{\sqrt{m}\,a_m^2}{\sqrt{n}(m+n)} < \pi \sum_{m=1}^\infty a_m^2,$$

and an analogous inequality holds for (b_n). Also, trivially,

$$\frac{a_m b_n}{m+n} = \frac{\sqrt[4]{m}\,a_m}{\sqrt[4]{n}\sqrt{m+n}} \cdot \frac{\sqrt[4]{n}\,b_n}{\sqrt[4]{m}\sqrt{m+n}},$$

so, by the Cauchy–Schwarz inequality,

$$\sum_{m,n} \frac{a_m b_n}{m+n} \le \left(\sum_{m=1}^\infty \sum_{n=1}^\infty \frac{\sqrt{m}\,a_m^2}{\sqrt{n}(m+n)}\right)^{1/2} \left(\sum_{n=1}^\infty \sum_{m=1}^\infty \frac{\sqrt{n}\,b_n^2}{\sqrt{m}(m+n)}\right)^{1/2}$$

$$< \pi \left(\sum_{m=1}^\infty a_m^2\right)^{1/2} \left(\sum_{n=1}^\infty b_n^2\right)^{1/2},$$

as claimed. $\qquad\qquad\square$

Notes. This inequality is due to David Hilbert, who proved it in his lectures with a larger constant than π; it was first reported in the dissertation of Hermann Weyl in 1908. The inequality with constant π, which later turned out to be best, was shown by Issai Schur. For sequences of bounded length, say, of length at

145

most a fixed value n, de Bruijn and Wilf came close to determining the best constant, which is (not surprisingly) smaller than π.

This particularly beautiful and short proof of Hilbert's inequality we have just given was published in 1993 by Krzysztof Oleszkiewicz, who also proved the following extension of Hilbert's inequality due to G.H. Hardy and M. Riesz. This extension was published in 1934 in Chapter 9 of the very influential monograph (entitled *Inequalities*) Hardy and Littlewood wrote with Pólya.

Let $p, q > 1$ be conjugate exponents, i.e. let $1/p + 1/q = 1$, and let (a_n) and (b_n) be sequences of non-negative reals such that $\sum a_n^p < \infty$ and $\sum_n b_n^q < \infty$. Then

$$\sum_{m,n=1}^{\infty} \frac{a_m b_n}{m+n} \leq \frac{\pi}{\sin(\pi/p)} \left(\sum_{m=1}^{\infty} a_m^p\right)^{1/p} \left(\sum_{n=1}^{\infty} a_n^q\right)^{1/q}.$$

References

de Bruijn, N.G. and H.S. Wilf, On Hilbert's inequality in n dimensions, *Bull. Amer. Math. Soc.* **68** (1962) 70–73.

Hardy, G.H., J.E. Littlewood and G. Pólya, *Inequalities*, Cambridge University Press (1952). Reprinted in paperback in Cambridge Mathematical Library (1988).

Oleszkiewicz, K., An elementary proof of Hilbert's inequality, *Amer. Math. Monthly* **100** (1993) 276–280.

Schur, I., Bemerkungen zur Theorie der beschränkten Bilinearformen mit unendlich vielen Veränderlichen, *J. Reine Angew. Math.* **140** (1911) 1–28.

43. The Central Binomial Coefficient

Let $k \geq 1$ be an integer and $c, d > 0$ positive real numbers such that

$$\frac{c}{\sqrt{k - 1/2}} 4^k \leq \binom{2k}{k} \leq \frac{d}{\sqrt{k + 1/2}} 4^k.$$

Then

$$\frac{c}{\sqrt{n - 1/2}} 4^n \leq \binom{2n}{n} \leq \frac{d}{\sqrt{n + 1/2}} 4^n \tag{1}$$

whenever $n \geq k$. In particular,

$$\binom{2n}{n} < \begin{cases} 2^{2n-1} & \text{if } n \geq 2, \\ 2^{2n-2} & \text{if } n \geq 5, \end{cases}$$

and

$$\frac{2^{2n-1}}{\sqrt{n - 1/2}} \leq \binom{2n}{n} \leq \frac{0.6}{\sqrt{n + 1/2}} 2^{2n}$$

for $n \geq 4$.

Proof. We prove (1) by induction on n, starting with the base case $n = k$ that holds by assumption. Assuming that (1) holds for n, let us prove it for $n + 1$. We have

$$\binom{2n + 2}{n + 1} = 2 \binom{2n + 1}{n} = 2 \left\{ \binom{2n}{n} + \binom{2n}{n - 1} \right\} = \frac{4(n + 1/2)}{n + 1} \binom{2n}{n}.$$

Hence, the two sides of (1) follow from

$$\frac{n + 1/2}{(n + 1)\sqrt{n - 1/2}} > \frac{1}{\sqrt{n + 1/2}}.$$

147

and

$$\frac{n + 1/2}{(n + 1)\sqrt{n + 1/2}} < \frac{1}{\sqrt{n + 3/2}},$$

both of which hold by the arithmetic mean–geometric mean (AM–GM) inequality or by simple expansions of the products. Indeed, it suffices to note that

$$(n + 1/2)^3 > (n + 1)^2(n - 1/2) \qquad \text{and} \qquad (n + 1/2)(n + 3/2) < (n + 1)^2.$$

Finally, to show that the constants 0.5 and 0.6 work for $n \geq 4$, note that

$$\frac{0.5}{\sqrt{3.5}} = 0.267 \cdots < \binom{8}{4} 4^{-4} = 0.273 \cdots < \frac{0.6}{\sqrt{4.5}} = 0.282 \cdots.$$

Evaluating the ratio $\binom{2n}{n} 4^{-n}$ for higher and higher values of n, the multiplicative constants in (1) can be brought closer and closer to each other, and approach $1/\sqrt{\pi} = 0.564189 \cdots$. ☐

Notes. The *Hint* we gave, 'Apply induction on n', is essentially content-free, as we are dealing with finite structures. Nevertheless, in this particular setup it does give some information: assume that our relations hold for a certain value of n, use those relations to deduce that it holds for the next value as well. This is exactly what we have done, in a mindless way.

A slightly more 'sophisticated' approach uses Stirling's formula, and gives us that

$$\binom{2n}{n} = \frac{1 + o(1)}{\sqrt{\pi n}} 4^{-n}.$$

This tells us that the multiplicative constants in (1) do tend to $1/\sqrt{\pi} = 0.564189 \cdots$. The advantage of the result we have proved over the more precise limiting result is that it can be used for all values of n without any further fuss.

44. Properties of the Central Binomial Coefficient

Let

$$\binom{2n}{n} = \prod_{p<2n} p^{\alpha_p}$$

be the prime factorization of the central binomial coefficient for $n \geq 1$, where p denotes a prime. Then:

(i) $\alpha_p = 0$ or 1 if $\sqrt{2n} < p < 2n$;
(ii) $\alpha_p = 0$ if $2n/3 < p \leq n$;
(iii) $p^{\alpha_p} \leq 2n$ for every p.

Proof. Throughout our proof, p denotes a prime number. Since the assertions are trivial for $n \leq 4$, we may assume that $n \geq 5$. For $2 \leq p \leq m$, the exponent of p in the prime factorization of $m!$ is

$$\lfloor m/p \rfloor + \lfloor m/p^2 \rfloor + \lfloor m/p^3 \rfloor + \cdots .$$

Hence, α_p, the exponent of $2 \leq p < 2n$ in the prime factorization of $(2n)!/(n!)^2$, is

$$\alpha_p = \sum_{k:p^k \leq 2n} \left(\lfloor 2n/p^k \rfloor - 2\lfloor n/p^k \rfloor \right). \tag{1}$$

Also, $\lfloor n/p^k \rfloor = \ell$ if and only if

$$\ell p^k \leq n < (\ell+1)p^k,$$

in which case

$$2\ell p^k \leq 2n < 2(\ell+1)p^k,$$

so $\lfloor 2n/p^k \rfloor = 2\ell$ or $2\ell + 1$. Consequently,

$$\lfloor 2n/p^k \rfloor - 2\lfloor n/p^k \rfloor = 0 \text{ or } 1.$$

The three assertions are easy consequences of these simple relations.

(i) Let $\sqrt{2n} < p < 2n$. Then $p^2 > 2n$, so there is only one summand in (1), and that summand is 0 or 1.

(ii) Let $2n/3 < p \le n$. As $n \ge 5$, we have $2n/3 > \sqrt{2n}$, so $p > \sqrt{2n}$. Furthermore, $2n/p < 2n/(2n/3) = 3$ so $\lfloor 2n/p \rfloor = 2$ and $2\lfloor n/p \rfloor = 2$ so $\alpha_p = 0$.

(iii) Let ℓ be the maximal integer such that $p^\ell \le 2n$. Then in the expression (1) for α_p there are exactly ℓ summands, and each of those summands is 0 or 1. Hence, $\alpha_p \le \ell$, and so $p^{\alpha_p} \le p^\ell \le 2n$.   □

45. Products of Primes

For a real number $n \geq 2$, denote by $\Pi(n)$ the product of all primes that are at most n. Thus, set

$$\Pi(n) = \prod_{p \leq n} p,$$

where, as usual, p stands for a prime. Then $\Pi(n) \leq 2^{2n-3}$ and, even more, $\Pi(n) < 4^n/n$ if $n \geq 9$.

Proof. Let us observe that if $p_1 = 2 < p_2 = 3 < \cdots$ is the sequence of primes and $p_k \leq x < p_{k+1}$ for some $k \geq 1$ then $\Pi(x) = \Pi(p_k)$: indeed, in the definition of $\Pi(x)$ we have exactly the same product as in that of $\Pi(p_k)$. In particular, if $m \geq 2$ is a natural number then $\Pi(2m) = \Pi(2m-1)$.

(i) Let us use induction on n to prove that $\Pi(n) < 2^{2n-3}$ for every natural number $n \geq 2$. This is trivially true for $n = 2$ (and 3, 5, 7, 11, say) and in the induction step we may take n odd: $n = 2m - 1$, say, where $m \geq 2$. Clearly, $\binom{2m-1}{m}$ is divisible by every prime p satisfying $m < p \leq 2m - 1$, so

$$\Pi(2m-1)/\Pi(m) \leq \binom{2m-1}{m} = \binom{2m-1}{m-1} = \frac{1}{2}\binom{2m}{m}.$$

Recalling from Problem 43 an upper bound on the central binomial coefficient $\binom{2m}{m}$, we find that

$$\frac{1}{2}\binom{2m}{m} < \frac{0.3}{\sqrt{m+1/2}} 4^m < 4^{m-1},$$

so, by the induction hypothesis,

$$\Pi(2m-1) < \Pi(m)4^{m-1} < 4^{2m-1}.$$

In fact, there is no need to recall that upper bound, since all we need is that

$$\binom{2m}{m} < 2^{2m-1} \tag{1}$$

for $m \geq 2$. This inequality holds for $m = 2$, since $6 < 8$, and for $m \geq 2$ we have

$$\binom{2m+2}{m+1} \Big/ \binom{2m}{m} = 2\binom{2m+1}{m} \Big/ \binom{2m}{m} = 2\frac{2m+1}{m+1} < 4,$$

so (1) follows by induction on m.

(ii) The inequality

$$\Pi(n) < 4^n/n,$$

which holds for $n \geq 9$, is almost as generous as the first, and is proved in a similar way, by induction on n. As before, we may take $n = 2m - 1$, where m is 'large', at least 5. Then

$$\Pi(2m-1) \leq \frac{1}{2}\binom{2m}{m}\Pi(m) < \frac{0.3}{\sqrt{m+1/2}}4^m \cdot 4^m/m < 4^{2m-1}/(2m-1),$$

completing the proof of the induction step.   □

Notes. The inequalities we have just proved are variations on a theme started by Paul Erdős in his beautiful proof of Bertrand's Postulate that we shall present in the next problem. Clearly, even the tighter inequality is *very* generous: using the upper bound in Problem 43, as we have just done, one can replace the factor $1/n$ by $n^{-c\log n}$ where $c > 0$; however, this extension is far from beautiful and is very far from a best possible result, so we shall not present it.

46. The Erdős Proof of Bertrand's Postulate

For every $n \geq 1$ there is a prime between n and $2n$: more precisely, there is a prime p satisfying $n < p \leq 2n$.

Proof. The assertion is that for every prime $p \geq 2$ there is another prime between p and $2p$. Equivalently, there is an infinite sequence of primes, $q_1 = 2 < q_2 < q_3 < \cdots$, say, such that $q_{k+1} < 2q_k$ for every k. Note that 2, 3, 5, 7, 13, 23, 43, 83, 163, 317, 631, 1259, 2503,... is the beginning of such a sequence, so in our proof we may assume that $n \geq 2503$. As we'll see, this is an overkill.

The young Erdős's idea of a proof was delightfully simple. Suppose n is large and there is no prime p satisfying $n < p \leq 2n$, i.e. $\Pi(2n)/\Pi(n) = 1$ where, as before, $\Pi(m)$ is the product of all primes that are at most m. Recalling the assertions from Problem 44 this tells us that $\binom{2n}{n}$ is rather 'small'. On the other hand, in Problem 43 we proved a lower bound on this central binomial coefficient, and for n large enough these bounds contradict each other.

These steps are essentially trivial to carry out. Suppose that the assertion fails for some n at least $2^9 = 512$ and so for this value of n we have $\Pi(2n)/\Pi(n) = 1$. Then, recalling the results from Problems 43 and 44, we find that

$$2^{2n-1}/\sqrt{n} < \binom{2n}{n} = \prod_{p \leq 2n} p^{\alpha_p} = \prod_{p \leq 2n/3} p^{\alpha_p} \leq \prod_{p \leq 2n/3} p \prod_{p \leq \sqrt{2n}} (2n), \quad (1)$$

and so, using an upper bound for $\Pi(2n/3)$ from Problem 45,

$$2^{2n-1}/\sqrt{n} < \Pi(2n/3) \prod_{p \leq \sqrt{2n}} (2n) \leq 2^{4n/3-3} (2n)^{\sqrt{2n}-1}. \quad (2)$$

Inequality (2) is trivially false if n is large enough. Indeed, setting $N = \sqrt{2n}$,

(2) implies that

$$2^{N/3} < N^2; \tag{3}$$

for $n = 2^9$, i.e. $N = 2^5$, inequality (2) is $2^{2^5/3} < 2^{10}$, a contradiction. Also, for $N \geq 2^5$ the derivative of the left-hand side of (2) is greater than that of the right-hand side, so (1) is false for all $n \geq 2^9$, completing the proof. $\square$

Notes. Bertrand's Postulate. The assertion in this problem that there is a prime number between a positive integer and its double was conjectured in 1845 by Joseph Bertrand, who also checked it up to some millions. This conjecture was proved by Pafnuty Lvovich Chebyshev in 1852. Since then numerous other proofs and generalizations have been published. In particular, in 1919 Ramanujan used some properties of the gamma function to give a simpler and shorter proof. In his very first 'real paper', published when he was nineteen, Paul Erdős gave a surprisingly simple and beautiful *elementary* proof: this is the proof we have just presented. In hindsight, this beautiful elementary proof is so 'natural' and simple that it is often presented in a first course on number theory without giving any credit to Erdős: to the question 'Which proof of Bertrand's Postulate have you given in your course?' the lecturer is likely to reply 'I gave the standard proof based on properties of the central binomial coefficient'. In fact, this beautiful proof did need the genius of the young Paul Erdős: *tanquam ex ungue leonem.*

Let us draw attention to the fact that in the version of this proof given above our attitude was rather cavalier: we could have made better use of the results in the preliminary problems. The reader may amuse himself by reducing the bound up to which the postulate has to be checked by hand: starting from inequality (1), one may use a better bound on $\Pi(2n/3)$ and a much better bound on $\pi(\sqrt{2n})$, the number of primes that are at most $\sqrt{2n}$, than the bound $\sqrt{2n} - 1$ we have used to find that $\prod_{p \leq \sqrt{2n}}(2n)$ is at most $(2n)^{\pi(\sqrt{2n})}$ rather than only $(2n)^{\sqrt{2n}-1}$. However, this reduction has no significance whatsoever: in the age of computers, checking that Bertrand's Postulate is true up to a pretty large number is not mathematics but an utter triviality.

References

Erdős, P., Beweis eines Satzes von Tschebyschef, *Acta Litt. Sci. Szeged* **5** (1932) 194–198.

Ramanujan, S., A proof of Bertrand's postulate, *J. Indian Math. Soc.* **11** (1919) 181–182. Reprinted in *The Collected Papers of Srinivasa Ramanujan*, pp. 208–209, AMS Chelsea Publ. (2000).

47. Powers of 2 and 3

No perfect powers of 2 and 3 differ by exactly 1, except 2 and 3, 4 and 3, and 8 and 9.

Proof. Clearly, for $n \leq 2$ the solutions of the equation $3^m = 2^n \pm 1$ are $(m, n) = (1, 1)$ and $(m, n) = (1, 2)$, giving $3 = 2 + 1$ and $3 = 4 - 1$, so our task is to show that for $n \geq 3$ the equation $3^m = 2^n \pm 1$ has only one solution, $(m, n) = (2, 3)$, giving $9 = 8 + 1$.

Taking the sequences $2, 4, 8, \ldots$ and $3, 9, 27, \ldots$ of powers of 2 and 3 modulo 8, we get the sequences

$$2, 4, 0, 0, 0, \ldots \qquad \text{and} \qquad 3, 1, 3, 1, 3, \ldots$$

Therefore, if $3^m = 2^n \pm 1$ and $n \geq 3$ then $m \geq 2$ is even so that 3^m is congruent to 1 modulo 8 and $3^m = 2^n + 1$. But if $m = 2k \geq 2$ is even then

$$3^m - 1 = 3^{2k} - 1 = (3^k - 1)(3^k + 1) = 2^n,$$

so $3^k - 1$ and $3^k + 1$ are perfect powers of 2 differing by 2. Hence, $3^k - 1 = 2$ and $3^k + 1 = 4$, implying that $m = 2$ and $n = 3$, giving $3^2 = 2^3 + 1$. $\square$

Notes. This lovely simple result was proved by the French Jewish scholar Gersonides (1288–1344), also known as Levi ben Gershom or Levi ben Gerson: it is likely to be the first result concerning a special case of Catalan's Conjecture, about which we shall say a few words in Problem 50. It appeared in 1343 in *De Numeris Harmonicis* ('The Harmony of Numbers'), a book Gersonides wrote on geometry at the request of Philip of Vitry, Bishop of Meaux. Gersonides was a medieval polyhistor – physician, astrologer, astronomer, mathematician and philosopher; later Jewish scholars criticized his unorthodox views, and some considered his commentary of the Old Testament heretical. In view of this little result being the first case of Catalan's famous conjecture, it is quaint that Gersonides was the son of Gerson ben Solomon *Catalan*.

48. Powers of 2 Just Less Than a Perfect Power

The only solution of $2^m = r^n - 1$ in positive integers m, r, n greater than 1 is $m = 3$, $r = 3$ and $n = 2$, giving $2^3 = 3^2 - 1$.

Proof. Let

$$r^n = 2^m + 1, \qquad \text{where } r, n, m > 1.$$

Our task is to show that $r = 3$, $n = 2$ and $m = 3$.

Let us start with the essentially trivial case when n is even, say, $n = 2k$. In this case,

$$r^{2k} - 1 = (r^k - 1)(r^k + 1) = 2^m,$$

so $r^k - 1$ and $r^k + 1$ are powers of 2 differing by 2. Hence $r^k - 1 = 2$, so $r = 3$, $n = 2k = 2$ and $m = 3$.

Turning to the heart of the assertion, suppose that there is a solution when n is odd, $n \geq 3$. Our task is to arrive at a contradiction. Clearly, r has to be odd, say, $r = 1 + 2^k q$, where $k \geq 1$ and q is odd. Then

$$r^n \equiv 1 + 2^k qn \quad \text{modulo } 2^{k+1},$$

so

$$2^m \equiv 2^k qn \quad \text{modulo } 2^{k+1}.$$

Since qn is odd, this tells us that $2^{m-(k+1)} - 1/2$ is an integer, so $m = k$ and

$$r = 1 + 2^m q \geq 1 + 2^m.$$

Consequently,

$$2^m + 1 = r^n \geq (1 + 2^m)^n \geq (1 + 2^m)^3 > 2^m + 1.$$

This contradiction completes our proof. $\qquad\qquad\square$

Notes. This is another special case of Catalan's Conjecture, together with Problem 49; it is quite a bit more general than Problem 47.

49. Powers of 2 Just Greater Than a Perfect Power

The equation $2^m = r^n + 1$ has no solutions in positive integers m, r, n greater than 1.

Proof. Suppose

$$r^n = 2^m - 1, \qquad \text{where } r, n, m > 1.$$

Our task is to arrive at a contradiction.

When n is even, this is trivial: since r is odd, $r^n \equiv +1$ modulo 4 and $2^m - 1 \equiv -1$ modulo 4, so they cannot be equal.

Assuming that $n \geq 3$ is odd, write r as $r = 2^k q - 1$, where $k \geq 1$, and q is odd. Then

$$r^n \equiv 2^k qn - 1 \quad \text{modulo } 2^{k+1},$$

so 2^{k+1} divides $2^m - 2^k qn$, i.e. $2^{m-(k+1)} - 1/2$ is an integer. Hence $m = k$ and so

$$r = 2^m q - 1 \geq 2^m - 1.$$

Consequently,

$$2^m - 1 = r^n \geq (2^m - 1)^n \geq (2^m - 1)^3 > 2^m - 1.$$

This contradiction completes our proof. □

Notes. This is yet another special case of Catalan's Conjecture. This result and the one in Problem 48 show that the only solution of the equation $r^n = 2^m \pm 1$ in integers greater than 1 is $r = 3$, $n = 2$ and $m = 3$, giving a considerable extension of the result of Gersonides in Problem 47, which concerned the case $r = 3$.

158

50. Powers of Primes Just Less Than a Perfect Power

For a prime $p \geq 3$, the equation $p^m = r^n - 1$ has no solution in positive integers m, r, n greater than 1.

Proof. Suppose

$$r^n - 1 = p^m, \qquad \text{where } r, n, m > 1,$$

where p is an odd prime. Our task is to arrive at a contradiction.

First, we know from Problem 49 that $r \neq 2$, so $r \geq 3$. Second, we may assume that n is a prime number, since if q is a prime factor of n then

$$\left(r^{n/q}\right)^q - 1 = p^n,$$

so replacing r by $r^{n/q}$ the exponent n is changed to q. We may write

$$r^n - 1 = (r - 1)(1 + r + r^2 + \cdots + r^{n-1}) = p^m,$$

so the two factors in the central product are perfect powers of p. Consequently, $r = p^s + 1$ for some $s \geq 1$, and so the second factor is not only a perfect power of p, but is also congruent to n modulo p^s. Therefore,

$$n \equiv 0 \pmod{p},$$

implying that $n = p$ and so

$$1 + r + \cdots + r^{p-1} = p^{m-s}.$$

Hence $m - s \geq 2$ and

$$p \equiv p^{m-s} \pmod{p^s},$$

telling us that $s = 1$ and so $r = p + 1$. Substituting $r = p + 1$ and $n = p$ into our starting equation, we find that

$$(p + 1)^p - 1 = p^m. \tag{1}$$

159

But this is impossible, because $(p + 1)^p - 1$ is not a perfect power of p, since

$$p^p < (p + 1)^p - 1 < p^{p+1}$$

for $p \geq 3$. Indeed, the first inequality (1) holds because $p^p < p^p + p^2 < (p + 1)^p - 1$, and the second because

$$(1 + 1/p)^p < e < p.$$

This contradiction completes our proof.   □

Notes. This is yet another special case of Catalan's Conjecture: rather than concerning the base 2, it concerns bases that are primes.

In 1826, August Leopold Crelle (1780–1855) founded a mathematics journal, the *Journal für die Reine und Angewandte Mathematik*, in Berlin which was not under the jurisdiction of an academy: this journal, commonly referred to as *Crelle's Journal* or just *Crelle*, is still going strong, as one of the most prestigious journals today. Inspired by this journal and Joseph Liouville's *Journal de Mathématiques Pures et Appliquées*, in 1842, Camille-Christophe Gérono (1799–1891) and Olry Terquem (1782–1862) founded a mathematics journal in France, *Nouvelles Annales de Mathématiques*, which went through several series till it ceased publication in 1927.

In addition to research papers (with only the surnames of the authors appearing, without their initials), this journal also published pedagogical articles, extracts of papers published elsewhere, and problems and conjectures. In the section headed *Theorems and Problems* of the first volume of this journal, Eugène Charles Catalan (1814–1894) published the following brief announcement:

> *Théoréme. Deux nombres entier consécutifs, autres que 8 et 9, ne peuvent être des puissances exactes.*

In fact, at the time Catalan had no proof of this result, neither did he manage to find one later, and eventually this came to be known as *Catalan's Conjecture*. Catalan soon realized that his 'theorem' was a great question, so two years later he republished it in the more prestigious *Crelle's Journal* in the form of a letter to the editor, still referring to his conjecture as a theorem:

Je vous prie, Monsieur, de vouloir bien énoncer, dans votre recueil,
le théorème suivant, que je crois vrais, bien que je n'aie pas encore
réussi à le démontrer complètement: d'autres seront peut-être plus
heureux:

Deux nombres entiers consécutifs, autres que 8 et 9, ne peuvent
être des puissances exactes; autrement dit: l'équation

$$x^m - y^n = 1, \tag{2}$$

dans laquelle les inconnues sont entières et positives, n'admet qu'une
seule solution.

Clearly, a 'puissance exact', an exact (or perfect) power, is meant to be a power with exponent at least 2.

Just about everyone now calls the equation above *Catalan's equation*, and the conjecture itself *Catalan's Conjecture*. After the very special case of Catalan's Conjecture proved by Gersonides and presented in Problem 47, in which the bases are 2 and 3, and its extensions in Problems 48 and 49, the first progress on this conjecture was made by Lebesgue – not the great mathematician best known for his measure-theoretic investigations, but Victor-Amédée Lebesgue (1791–1875) – who in 1850 proved that $x^m = y^2 + 1$ has no solution in positive integers. To an undergraduate today, Lebesgue's proof would be as expected: it uses factorization in $\mathbb{Z}[i] = \mathbb{Z}[\sqrt{i}]$, the ring of Gaussian integers. Then, in 1870 and 1871, Gérono went much further: he proved that (2) has no solution in positive integers if either x or y is a prime number. The case when y is a prime is the content of this problem.

It is a great pleasure to give credit to Paul Erdős, who in the early 1960s told me about these results of Gérono: I cannot overestimate how terribly impressed I was that he could recall two papers that had been published almost one hundred years earlier. Today this would be much less impressive, but that was well before the era of the internet: the only way he would have known about Gérono's theorem is if he had read the relevant papers in a library. I suspect (only now!) that (as an undergraduate?) he had made an effort to look up Gérono's results because he himself was trying to prove Catalan's Conjecture.

Let us note that the second half of Gérono's result, that $x = 3$, $m = 2$, $y = 2$ and $n = 3$ is the only solution of (2) in integers greater than 1 if x is a prime, easily follows from the Lebesgue theorem we have just mentioned, and the method Gérono applied to prove the result in this problem, i.e. Catalan's Conjecture in the case when y is a prime. In fact, this is the result Gérono published first, in 1870. We prove it here, starting with a statement of the theorem in the notation we have used.

Let $p \geq 3$ be an odd prime and $p^m = r^n + 1$, where m, r and n are integers greater than 1. Then $p = 3$, $m = 2$, $r = 2$ and $n = 3$.

In proving this, as above, we may assume that n is a prime; also, by Lebesgue's theorem above that $n \neq 2$ we see that $n \geq 3$, so n is an odd prime. As n is odd, $r^n + 1$ can be factorized, giving us

$$p^m = (r + 1)(r^{n-1} - r^{n-2} + r^{n-3} - \cdots - r + 1).$$

As the two factors must be perfect powers of p, we see that $r + 1 = p^s$ for some $s \geq 1$, and so the second factor is not only a perfect power of p but is also congruent to n modulo p^s. Hence n is a multiple of p, so $n = p \geq 3$, implying that $s = 1$ and so $p = r + 1$. Therefore our equation becomes

$$p^m = (p - 1)^p + 1.$$

For $p = 3$ we find the solution $m = 2$, $r = 2$ and $n = 3$, giving us the neighbouring perfect powers 8 and 9 of Catalan's Conjecture. However, for $p \geq 5$ this equation has no solution in a positive integer m since

$$p^{p-1} < (p - 1)^p + 1 < p^p,$$

so the right-hand side is not a perfect power of p. Indeed, the second inequality is trivial, and the first holds since for $p \geq 5$ we have

$$1 < \frac{p - 1}{e} < (1 - 1/p)^{p-1}(p - 1) = (p - 1)^p / p^{p-1}.$$

Not surprisingly, the first three contributions to Catalan's Conjecture, one by Lebesgue and two by Gérono, were published in *Nouvelles Annales de Mathématiques*, the journal of Gérono and Terquem, where Catalan first published his 'theorem'. However, after these results, Catalan's Conjecture faded from mathematics for a long time. It re-emerged in the 1950s when LeVeque and then Cassels wrote papers about it: mistakenly, LeVeque attributed the start of Catalan-type questions to a paper Pillai wrote in 1931.

After the papers of Cassels, more and more work was done on Catalan's Conjecture; in particular, Baker's fundamental results on linear forms in logarithms proved in 1964 had a great influence on this work. In 1976 Tijdeman made use of Baker's 'sharpening' results to prove that there is an effectively computable bound on the size of the exponents in Catalan's equation. The first explicit bound was of the order 10^{110}; eventually this was reduced to less than 10^{16}. And then in 2002 came the great news that Preda Mihăilescu had given a full proof of the conjecture. It was most appropriate that the great paper of Mihăilescu appeared, in 2004, in *Crelle's Journal*, the high-class journal where Catalan republished his 'theorem'. We hasten to emphasize that the elementary arguments we have used to prove some simple cases of Catalan's Conjecture have nothing to do with the high-class proofs of Tijdeman and Mihăilescu. To conclude our remarks about Catalan's Conjecture, we reproduce the abstract of Mihăilescu's paper, although only mathematicians are likely to understand most of it.

Catalan's conjecture states that the equation $x^p - y^q = 1$ has no other integer solutions [in integers greater than 1] but $3^2 - 2^3 = 1$. A classical result of Cassels and our recent consequence establish that p, q must verify a double Wieferich condition if the equation has integer solutions with p, q odd. If $p \equiv 1 \bmod q$, then a contradiction to the above follows from Baker's methods in transcendence theory. If $p \not\equiv 1 \bmod q$, then the Galois group of $\mathbb{Q}(\zeta)/\mathbb{Q}$ has order coprime to q. We show that the existence of solutions to Catalan's equation produces an excess of q-primary cyclotomic units in this case. This fact leads to a contradiction which proves Catalan's conjecture.

References

Baker, A., Linear forms in the logarithms of algebraic numbers I, II, III, *Mathematika* **13** (1966) 204–216; **14** (1967) 102–107; and **14** (1967) 220–228.

Baker, A., A sharpening of the bounds for linear forms in logarithms I, II, *Acta Arith.* **21** (1972) 117–129; **24** (1974) 33–36; and **27** (1975) 247–252.

Cassels, J.W.S., On the equation $a^x - b^y = 1$, *Amer. J. Math.* **75** (1953) 159–162.

Cassels, J.W.S., On the equation $a^x - b^y = 1$ II, *Proc. Cambridge Philos. Soc.* **56** (1960) 97–103.

Ko, C., On the Diophantine equation $x^2 = y^n + 1$, $xy \neq 0$, *Sci. Sinica* **14** (1965) 457–460.

LeVeque, W. J., On the equation $a^x - b^y = 1$, *Amer. J. Math.* **74** (1952) 325–331.

Mihăilescu, P., Primary cyclotomic units and a proof of Catalan's conjecture, *J. Reine Angew. Math.* **572** (2004) 167–195.

Pillai, S.S., On the inequality $0 < a^x - b^y \leq n$, *J. Indian Math. Soc.* **19** (1931) 1–11.

Tijdeman, R., On the equation of Catalan, *Acta Arith.* **29** (1976) 197–209.

51. Banach's Matchbox Problem

An inveterate smoker puts two boxes of matches into the pockets of his jacket. Every time he wants to light up, he is equally likely to reach into either pocket. After a while, when he takes out one of the boxes, he finds it empty. Having started with n matches in each box, the probability that at that moment the other box has k matches for some k with $0 \le k \le n$ is

$$\binom{2n-k}{n-k}2^{k-2n}.$$

Proof. Let us change the setup: let us assume that each box has an infinite number of matches. The smoker chooses $2n - k + 1$ matches at random, and marks the box of the last match chosen A, and the other B. Then the probability in question is exactly the probability that of the first $2n - k$ matches n come from A. All sequences $ABBA \ldots$ are equally likely, so every sequence of length $2n - k$ has probability $(1/2)^{2n-k}$. As there are $\binom{2n-k}{n}$ 'desirable' sequences, the probability is

$$\binom{2n-k}{n-k}2^{k-2n},$$

as claimed.   □

Notes. This is a truly old chestnut, and should be known by all mathematicians who have taken an introductory course in probability theory.

According to William Feller, the attribution of this problem to the famous Polish analyst, Stefan Banach, is incorrect. Although Banach *was* an inveterate smoker, the problem was inspired by a speech Hugo Steinhaus, another well-known Polish mathematician, gave in honour of Banach.

Appropriately, I heard the problem from Alfréd Rényi, the great Hungarian probabilist, when he gave an introductory course on probability. Rényi him-self was an inveterate smoker, and he illustrated the problem with his own matchboxes. Tragically, Rényi died of lung cancer in 1970, aged 48.

A usual consequence of this 'result' is the following beautiful binomial identity:

$$\binom{2n}{n} + 2\binom{2n-1}{n} + 2^2\binom{2n-2}{n} + \cdots + 2^n\binom{n}{n} = 2^{2n}.$$

To see this, simply note that when the matchbox is found to be empty, the other box has k matches for some k between 0 and n. Hence, the sum of the probabilities of these events is 1:

$$2^{-2n}\binom{2n}{n} + 2^{-2n+1}\binom{2n-1}{n} + \cdots + 2^{-n}\binom{n}{n} = 1.$$

Multiplying this by 2^{2n}, we get our identity.

52. Cayley's Problem

For $3 \le k < 2k \le n$, write $f(n,k)$ for the number of convex k-gons inscribed in a convex n-gon in such a way that every side of the k-gon is a diagonal of the n-gon. Then

$$f(n,k) = \frac{n}{n-k}\binom{n-k}{k}.$$

Proof. Let $x_1 x_2 \cdots x_n$ be our n-gon and let us count the number of k-gons containing x_1. This is the number of $(k-1)$-subsets of the vertex set $\{x_3, x_4, \dots, x_{n-1}\}$ containing no two neighbours, which is $\binom{n-k-1}{k-1}$. Indeed, there is a one-to-one correspondence between the $(k-1)$-subsets $\{i_1, \dots, i_{k-1}\}$ of $\{3, 4, \dots, n-k+1\}$ with $i_1 < \cdots < i_{k-1}$ and the collection of vertex sets $\{x_{i_1}, x_{i_2+1}, x_{i_3+2}, \dots, x_{i_{k-1}+k-2}\}$ of vertices without neighbours. Finally, we have n choices for a vertex of the n-gon and k choices for a vertex of our inscribed k-gon, so

$$f(n,k) = \frac{n}{k}\binom{n-k-1}{k-1} = \frac{n}{n-k}\binom{n-k}{k},$$

as required. $\qquad\square$

Notes. The widely accepted name of this problem is 'Cayley's problem'. I have used it with some reluctance, since it seems to be wrong to associate the name of the great Arthur Cayley with such a simple problem.

A natural extension of this problem is to inscribe polygons formed by 'long' diagonals. Define the *length* of a diagonal xy of a convex n-gon to be the graph distance between x and y, so that the length of a diagonal $x_i x_{i+j}$, $1 \le i < i + j \le n$, of a convex n-gon $x_1 x_2 \cdots x_n$ is $\min\{j, n-j\}$, and write $f(n, k; \ell)$ for the number of convex k-gons formed by diagonals of length at least $\ell + 1$ of

166

a convex n-gon. Thus $f(n, k; 1) = f(n, k)$. With minimal changes in the proof above, we find that

$$f(n, k; \ell) = \frac{n}{n - k\ell} \binom{n - k\ell}{k}.$$

In fact, the proof we have given is rather pedestrian. For a less pedestrian proof of the more general assertion, given a k-subset $\{i_1, \ldots, i_k\}$ of $\{1, \ldots, n - k\ell\}$ with $i_1 < \cdots < i_k$, map it into a k-subset of the convex n-gon $x_1 \cdots x_n$, with the indices taken modulo n, as follows. Map i_1 into a vertex j_1, then i_2 into $j_2 = j_1 + (i_2 - i_1) + \ell$, i_3 into $j_3 = j_2 + (i_3 - i_2) + \ell$, etc. This gives a convex k-gon formed by long diagonals, and every such k-gon is obtained in this way. Finally, there are n choices for j_1, but every k-gon is obtained $(n - k\ell)$ times, proving our assertion.

53. Min vs Max

Let K_n be a complete graph of order n with positive weights on the edges. Define the weight of a subgraph to be the sum of the weights of its edges. Consider two 'natural' ways of constructing a Hamilton path (H-path) of K_n. First, starting at a vertex a, always choose the edge of maximal weight continuing the path constructed so far, joining the current end-vertex to a vertex not on the current path. Second, starting from a vertex b, always choose the edge of minimal weight. Then the weight of the first H-path is at least as large as the weight of the second.

Proof. The condition that the weights are non-negative is a red herring, as we may add a constant to make all the weights positive.

Also, we may assume that each weight is 0 or 1. Indeed, given real numbers $x_1, \ldots, x_n$ and $y_1, \ldots, y_n$, if, for every z, the number of x_i greater than z is at most the number of y_i greater than z, then $\sum_{i=1}^{n} x_i \leq \sum_{i=1}^{n} y_i$. Assuming, as we may, that $x_1 \leq \cdots \leq x_n$, this condition actually implies (rather trivially) that $x_i \leq y_i$ for every i, with $1 \leq i \leq n$. In view of this, our assertion follows from the following claim we state for colours, rather than 0–1 weights.

Claim *Consider two algorithms for constructing an H-path in the complete graph K_n whose edges are coloured red and blue. In the 'red algorithm', grow a path from a vertex a by choosing a red edge, whenever possible; analogously, in the 'blue algorithm', start from a vertex b, and choose a blue edge whenever possible. Let H_r and H_b be the red–blue coloured H-paths constructed in this way. Then H_r has at least as many red edges as H_b.*

Proof of Claim Let $H_b = x_1 x_2 \ldots x_n$, where $x_1 = b$, and let $R_1, \ldots, R_k$, $R_i = x_{\ell_i} x_{\ell_i+1} \ldots x_{m_i}$, be maximal red paths on H_b, with vertex sets $S_i = \{x_{\ell_i}, x_{\ell_i+1}, \ldots, x_{m_i}\}$ with $s_i = |S_i|$ vertices. Thus for $\ell_i \leq j < m_i$ the edge

168

$x_j x_{j+1}$ is red, and the edges $x_{\ell_i-1} x_{\ell_i}$ and $x_{m_i} x_{m_i+1}$ are blue, provided they exist. Note that H_b has $\sum_{i=1}^{k}(s_i - 1)$ red edges.

Since R_i is a part of a 'blue' H-path, for $\ell_i \le u < m_i$ and $u < v \le n$ the edge uv is red, for otherwise in H_b the path $x_1 \ldots x_u$ could have been continued with the blue edge $x_u x_v$, so it would not have been continued with the red edge $x_u x_{u+1}$. In particular, every edge joining two vertices of S_i is red.

Now, let $H_r = y_1 y_2 \ldots y_n$, where $y_1 = a$, be our 'red' H-path. Then for each i, $1 \le i \le k$, the first $s_i - 1$ vertices of S_i occurring on H_r are followed by a red edge in H_r. Indeed, when we chose an edge incident with one of these $s_i - 1$ vertices, we could have gone to a vertex of S_i not yet on H_r, and that edge is red. Hence the current path could have been continued with a red edge, so the red algorithm did choose a red edge, proving our claim. Consequently, H_r has at least $s_i - 1$ red edges *starting* at a vertex belonging to S_i. In particular, H_r has at least $\sum_{i=1}^{k}(s_i - 1)$ red edges. This completes the proof of the Claim and so the assertion of the problem. □

Note. I heard this problem in December 2016 from Ernst Fischer, who at the time was a student in a Gymnazium.

54. Sums of Squares

If any three numbers are taken that cannot be arranged in an arithmetic progression, and whose sum is a multiple of 3, then the sum of their squares is also the sum of another set of 3 squares, the two sets having no common term.

Proof. Let a, b and c be our three numbers, and note the following identity:

$$(a^2 + 4b^2 + 4c^2) + (4a^2 + b^2 + 4c^2) + (4a^2 + 4b^2 + c^2) = 9(a^2 + b^2 + c^2).$$

Hence,

$$a^2 + b^2 + c^2 = \frac{1}{9}\{(a^2 + 4b^2 + 4c^2 + 8bc - 4ca - 4ab)$$
$$+(4a^2 + b^2 + 4c^2 - 4bc + 8ca - 4ab)$$
$$+(4a^2 + 4b^2 + c^2 - 4bc - 4ca + 8ab)\}$$
$$= \left(\frac{-a + 2b + 2c}{3}\right)^2 + \left(\frac{2a - b + 2b}{3}\right)^2 + \left(\frac{2a + 2b - c}{3}\right)^2$$
$$= A^2 + B^2 + C^2.$$

Since $a + b + c$ is a multiple of 3, so are $-a + 2b + 2c$, $2a - b + 2c$ and $2a + 2b - c$, so A, B and C are integers, the sum of whose squares is $a^2 + b^2 + c^2$.

To complete our proof, we have to show that the sets $\{a, b, c\}$ and $\{A, B, C\}$ have no common element. Suppose for a contradiction that they do share an element: we may assume that this is $A = (-a + 2b + 2c)/3$. There are three possibilities: $A = a$ or $A = b$ or $A = c$ – each of these leads to the contradiction that the elements a, b, c can be arranged in an arithmetic progression. Indeed, $A = a$ tells us that $a = (b + c)/2$, $A = b$ gives that $c = (a + b)/2$, and $A = c$ that $b = (a + c)/2$, completing our proof. $\square$

Notes. The Reverend Charles Ludwidge Dodgson (1832–1898), better known as the comic genius Lewis Carroll, was the mathematics lecturer of Christ

Church, the great Oxford college. This is Problem 61 of Charles Dodgson's (i.e. Lewis Carroll's) collection of problems.

Reference

C.L. Dodgson, *Curiosa Mathematica, Part II, Pillow Problems, Thought out During Wakeful Hours*, 2nd edn, Macmillan (1893).

55. The Monkey and the Coconuts

Five men and a monkey were shipwrecked on a desert island. They spent the first day gathering coconuts, which they piled up in a heap before going to sleep. The heap was very big, but could not possibly have contained more than ten thousand coconuts.

In the middle of the night one of the men woke up, and to make sure that he was not shortchanged, divided the coconuts into five equal piles, with a single coconut remaining. He gave that remainder to the monkey, hid his fifth, rearranged the rest into a single heap, and went back to sleep. Later a second man woke up and did the same, then the third, the fourth and the fifth. In the morning the men successfully divided the remaining coconuts into five equal piles: this time no coconuts were left over.

Under these conditions, in the beginning there must have been 3,121 coconuts.

Solution. Write N_i for the number of coconuts left after the visit of the ith man, so that in the beginning there were N_0 coconuts, and N_5, the number of coconuts in the morning, is a multiple of 5. The rearrangements of the coconuts tell us that $N_{i+1} = \frac{4}{5}(N_i - 1)$ for $i = 0, \ldots, 4$. Hence, writing N_0 in the form $N_0 = -4 + 5^5 R$, we find that $N_i = -4 + 4^i 5^{5-i} R$ for $i = 1, \ldots, 5$.

To find the answer, we have to make sure that our conditions are satisfied: (i) each N_i is an integer, (ii) N_5 is a multiple of 5, and (iii) $N_0 \leq 10,000$.

First, as $N_0 + 4 = 5^5 R$ and $N_5 + 4 = 4^5 R$ are integers, R itself is an integer. Next,

$$N_5 = -4 + 4^5 R \equiv -4 - R \pmod{5},$$

i.e. R is congruent to 1 modulo 5, so its possible values are $1, 6, 11, \ldots$.
Finally,

$$N_0 = -4 + 5^5 R = -4 + 3125R \leq 10,000,$$

172

so R has to be 1, implying that at the beginning there were $N_0 = -4 + 3125 = 3121$ coconuts. □

Notes. As every mathematician knows, this puzzle is a simple Diophantine problem, a system of equations to be solved in integers. A somewhat (considerably?) simpler version of it does not ask for the final pile to be divisible into five equal parts. Almost sixty years ago, Paul Dirac, the great physicist, asked me to answer this simpler version while waiting for his wife to prepare our afternoon tea: this is the reason why this simple puzzle has made it into this collection. Dirac loved the fact that 'minus four' was a valid answer to this simple variant.

As I know very little about puzzles, it is only recently that I came across Singmaster's most informative paper about this puzzle and its long history. As Singmaster discovered, *The Monkey and the Coconuts* puzzle became notorious in 1926 when, in the 9th October issue of *The Saturday Evening Post*, Ben Ames Williams published a short story entitled 'Coconuts', which centred around it. Singmaster described the rich history of the problem going back several hundred years.

In fact, this problem has Oxbridge connections as well: Lewis Carroll, W.W. Rouse Ball and J.H.C. Whitehead all took an interest in it – in fact, Dirac heard the problem from Whitehead.

Reference

Singmaster, D., Coconuts: The history and solutions of a classic Diophantine problem, *Gaṇita-Bharati* **19** (1997) 35–51.

56. Complex Polynomials

Given a polynomial h with complex coefficients, write $S_h \subset \mathbb{C}$ for the region where $|h(z)| \leq 1$. If f and g are monic polynomials of degree at least one with $f \neq g$, then S_f cannot be a proper subset of S_g.

Proof. We may assume that f and g have the same degree, since if $\deg f = n$ and $\deg g = m$ then $S_f = S_{f^m}$, $S_g = S_{g^n}$ and $\deg f^m = \deg g^m = nm$.

We shall show a little more than claimed: if $\deg f = \deg g \geq 1$ and $S_g \subset S_f$ then $f = g$. To this end, we write U_f for the unbounded component of $\mathbb{C} \setminus S_f$, and define U_g analogously. Then $U_f \subset U_g$,

$$\varphi(z) = f(z)/g(z)$$

is analytic in U_f and $|\varphi(z)| \leq 1$ on the boundary ∂U_f of U_f. Since $\varphi(z) \to 1$ as $z \to \infty$, the maximum modulus theorem tells us that $\varphi(z)$ is identically 1 in U_f. This implies that f and g are identical, as claimed. □

Notes. If, as before, f and g are monic polynomials with $\deg f = n$ and $\deg g = m$ then we need not replace them by f^m and g^n: it suffices to replace them by $f^{N/n}$ and $g^{N/m}$, where N is the least common multiple of n and m. Our proof above tells us that these two polynomials are identical: $f^{N/n} = g^{N/m}$. But N/n and N/m are relatively prime, so the uniqueness of factorization over the complex numbers tells us that $f = h^{N/m}$ and $g = h^{N/n}$ for some monic polynomial h. Thus $S_g \subset S_f$ implies that f and g are integral powers of the same polynomial; in particular, $S_f = S_g$.

This little result is one of many that Paul Erdős presented in a mini-course(!) he gave at the Mathematical Institute in Budapest in 1964.

57. Gambler's Ruin

Rosencrantz and Guildenstern (Figure 35) gamble by repeatedly tossing a biased coin with probability p of heads and probability q = 1 − p of tails, where 0 < p < 1. Every time the coin comes up heads, Rosencrantz wins a krone from Guildenstern, otherwise Guildenstern wins a krone from Rosencrantz: they play till one of them 'gets ruined' and so the other 'wins'. Then the expected duration of the game is independent of who wins it.

Proof. For $p = 1/2$ there is nothing to prove, so we assume that $p \neq 1/2$.

Let us introduce two random variables, T and W. We write T for the duration of the game (i.e. the 'time' or the 'number of tosses' the game takes), and set $W = 1$ if Rosencrantz wins and $W = 0$ if Guildenstern does. Our task is to prove

Figure 35 Rosencrantz and Guildenstern. Sketch by Moyr Smith of a performance, published in *Black and White* in 1891.

that the expectation of T conditional on $W = 1$ is the same as the expectation of T conditional on $W = 0$:

$$\mathbb{E}(T \mid W = 1) = \mathbb{E}(T \mid W = 0).$$

In fact, we shall prove more than is required: we shall prove that the random variables T and W are independent:

$$\mathbb{P}(T = t \text{ and } W = w) = \mathbb{P}(T = t)\,\mathbb{P}(W = w) \tag{1}$$

for all t and w.

Now, there are several conditions equivalent to (1); perhaps the simplest is that

$$\mathbb{P}(T = t \mid W = 1) = \mathbb{P}(T = t \mid W = 0)$$

for every t. A slightly less obvious condition for independence will be useful:

$$\mathbb{P}(W = 1 \text{ and } T = t) = c\,\mathbb{P}(T = t) \tag{2}$$

for some constant c. To see that (2) implies (1), note that by (2) we have

$$\mathbb{P}(W = 1) = \sum_t \mathbb{P}(W = 1 \text{ and } T = t) = c \sum_t \mathbb{P}(T = t) = c.$$

Although this is clearly sufficient for the independence of T and W, i.e. for (1), let us spell out the case $W = 0$ as well:

$$\mathbb{P}(W = 0 \text{ and } T = t) = \mathbb{P}(T = t) - \mathbb{P}(W = 1 \text{ and } T = t)$$
$$= \mathbb{P}(T = t)(1 - \mathbb{P}(W = 1) = \mathbb{P}(W = 0)\mathbb{P}(T = t).$$

After these rather obvious remarks about the independence of random variables, let us turn to the specific random variables T and W we have defined. Let $R(t)$ and $G(t)$ be the sets of ways in which Rosencrantz and Guildenstern win in exactly t tosses, so that $R(t))$ is precisely the event that $W = 1$ *and* $T = t$. In particular,

$$\mathbb{P}(W = 1 \text{ and } T = t) = \mathbb{P}(R(t)).$$

Note that, starting with $k = 3$ kroner, $THTTHTT$ is one of the ways Guildenstern wins (and so Rosencrantz gets ruined) in $t = 7$ steps. Clearly, there are $|R(7)| = \binom{7}{2} = 21$ ways in which Rosencrantz can be ruined in seven steps. Keeping to $k = 3$ kroner, but changing the time from seven to nine, $THTHTTHTT$ is one of the ways Rosencrantz can be ruined in nine tosses: the sequence $HHHTTTTTT$ of nine tosses of three heads and six tails does not arise since Guildenstern would be ruined after the first three tosses.

In fact, we do not care about the cardinalities of $R(t)$ and $G(t)$: all we need is that every element of $R(t)$ is a string of $(t + k)/2$ heads H and $(t - k)/2$ tails T arranged in such a way that neither Rosencrantz nor Guildenstern gets ruined before t tosses are completed. Clearly, the same holds for $G(t)$, with H and T interchanged. This description tells us that by interchanging H and T, we set up a one-to-one correspondence between $R(t)$ and $G(t)$. Also, when changing a string in $G(t)$ to a string in $R(t)$, we increase the number of heads by k and decrease the number of tails by k. In particular, $R(t)$ and $G(t)$ have the same number of strings, and

$$\mathbb{P}(R(t)) = (p/q)^k\, \mathbb{P}(G(t)).$$

Hence,

$$\mathbb{P}(T = t) = \mathbb{P}(R(t)) + \mathbb{P}(G(t)) = \mathbb{P}(R(t))\left(1 + (q/p)^k\right),$$

and

$$\mathbb{P}(W = 1 \text{ and } T = t) = \mathbb{P}(R(t)) = \mathbb{P}(T = t)/\left(1 + (q/p)^k\right),$$

proving (2), and so completing the proof of our assertion. $\qquad\square$

Notes. The 'Gambler's Ruin' is one of the classical problems discussed in just about every introduction to probability theory, but the particular aspect of it appearing in this problem seems to have been first published by Stern in 1975. The elegant proof above was published by Samuels.

The proof above gives more than has been explicitly stated. Indeed, we have shown that

$$\mathbb{P}(T = t) = \mathbb{P}\left(1 + (q/p)^k\right)$$

and so

$$\mathbb{P}(R(t)) = \mathbb{P}(W = 1)\mathbb{P}(T = t) = \mathbb{P}(W = 1)\mathbb{P}(R(t))\mathbb{P}\left(1 + (q/p)^k\right).$$

Therefore the probability that Rosencrantz wins (and so Guildenstern is ruined) is

$$\mathbb{P}(W = 1) = 1/\left(1 + (q/p)^k\right) = \frac{(p/q)^k}{1 + (p/q)^k}.$$

Moreover, although we have not found the common value of $\mathbb{E}(T \mid W = 1) = \mathbb{E}(T \mid W = 0)$, as this equation holds for every t, the conditional expectations of T are also equal, and so they are equal to the unconditional expectation $\mathbb{E}(T)$ as well. This can be computed with martingales to tell us that

$$\mathbb{E}(T \mid W = 1) = \mathbb{E}(T \mid W = 0) = \frac{k}{q - p} \cdot \frac{1 - (p/q)^k}{1 + (p/q)^k}$$

if $p \neq 1/2$ (i.e. $p \neq q$), and

$$\mathbb{E}(T \mid W = 1) = \mathbb{E}(T \mid W = 0) = k^2$$

if $p = q = 1/2$.

The results of Stern and Samuels have been extended by many people, incuding Beyer and Waterman, Lengyel, and Gut.

References

Beyer, W.A. and M.S. Waterman, Symmetries for conditioned ruin problems, *Math. Mag.* **50** (1977) 42–45.

Gut, A., The gambler's ruin problem with delays, *Statist. Probab. Lett.* **83** (2013) 2549–2552.

Lengyel, T., The conditional gambler's ruin problem with ties allowed, *Appl. Math. Lett.* **22** (2009) 351–355.

Samuels, S.M., The classical ruin problem with equal initial fortunes, *Math. Mag.* **48** (1975) 286–288.

Stern, F., Conditional expectation of the duration in the classical ruin problem, *Math. Mag.* **48** (1975) 200–203.

58. Bertrand's Box Paradox

There are three identical boxes, with identical drawers on opposite sides, and a coin in each drawer. One of the boxes contains two gold coins, another two silver coins, and the third one of each. We pick a drawer at random, and find a gold coin. Then the probability that the other coin in the box is also gold is $2/3$.

Proof. Let us write G_1, G_2, S_1, S_2, G and S for the coins in the various boxes, with the obvious convention that G_1 and G_2 are the two gold coins in the two drawers of the gold–gold box, G and S are the gold and silver coins in the same box, and S_1, S_2 are the two silver coins in the same box. As we have to pick a gold coin, there are three possibilities: we may pick G_1, G_2 or G. In two of these three cases, the other coin is also gold, so the probability that the other coin in the box is also gold is indeed $2/3$. □

Notes. This very simple 'paradox' has been included to draw attention to the fact that the development of probability theory lagged behind that of the mainstream branches of mathematics such as analysis, geometry, algebra and number theory: the question above was posed by Joseph Bertrand in 1889.

Why is this a paradox? Because the following argument gives a different answer. There are three boxes, only two of which may be chosen: G_1G_2 and GS. Of these two boxes, the first will result in a gold coin, and the second in a silver. Hence the probability that the other coin is gold is 1/2. It is clearly best to leave the resolution of this paradox to the reader.

This question can be generalized in a number of ways. Thus we may have many drawers in each box (preferably the same number), and coins made of several alloys. A simple form of this is the following. *There are n + 1 boxes, each with n drawers, with a coin in each drawer. For every k, $0 \leq k \leq n$, there is a box with k gold coins and n − k silver coins. As before, we pick a coin at* *random, and find that it is gold. Then we pick another coin from the same box. What is the probability that that coin is also gold?*

In this version there are $n(n + 1)$ coins, half of which are gold and half are silver, so the probability that any particular coin is picked is $2/n(n + 1)$. The probability that we pick a gold coin from a box with k gold coins is $2k/n(n+1)$. Having picked a gold coin from a box with k gold coins, the probability that the other is also gold is $(k - 1)/(n - 1)$. Hence the probability that the second coin is also gold is

$$\sum_{k=0}^{n} \frac{2k}{n(n + 1)} \cdot \frac{k - 1}{n - 1} = \frac{2}{3}\binom{n + 1}{3}^{-1} \sum_{k=0}^{n} \binom{k}{2} = \frac{2}{3},$$

exactly as in the case $n = 2$.

Reference

Bertrand, J., *Calcul des Probabilités*, Gauthier-Villars (1889).

59. The Monty Hall Problem

There are three doors, one of which conceals a car and the others a goat. The contestant is told that eventually he will open a door and win whatever is behind that door.

First, Monty asks the contestant to choose a door but not open it. The contestant chooses door A. Then Monty, who knows perfectly well where the car is, chooses another door, say, B, knowing that it conceals a goat, and opens it. After this, Monty offers the contestant the chance of switching from his original guess, A, to the other unopened door C. What should the contestant do to optimize his chances of winning the car? Should he swap or stay with his original choice?

Solution. By switching, the contestant will double his chances of winning the car: the probability will increase from $1/3$ to $2/3$. He should definitely switch.

To see this, let us call the door chosen by the contestant door A. First, suppose that the contestant guesses correctly, and door A hides the car. The probability of this is $1/3$. Hence, by staying with his choice, his probability of winning the car is $1/3$.

Second, suppose that the contestant guesses incorrectly, and a goat is behind door A. The probability of this is $2/3$. By swapping, *he wins the car* since the car has to be behind one of the two other doors, and Monty reveals which of them does *not* conceal a car. Hence, by swapping, the contestant wins the car with probability $2/3$. □

Notes. This rather simple puzzle became notorious in 1990 when, in her column entitled 'Ask Marilyn' in *Parade Magazine*, Marilyn vos Savant responded politely to a reader's inquiry about this problem. Although the 'solution' provided by vos Savant was correct, she received over 10,000 letters, many from academics, even mathematicians, rudely stating that she was totally wrong. It is because of this shameful episode, showing up academics, especially mathematicians, in a very bad light that this puzzle is included in the present collection. Another reason is that, as I have found, even some Fellows of Trinity College were puzzled by the correct solution.

The reader may have noticed that the formulation of the problem in the list of problems is pretty different from its formulation appearing among the solutions, more so than in any other case. This is simply because the first explicit occurrence of the Monty Hall Problem (MHP) was in 1975, when the statistician Steve Selvin wrote it up in the 'Letters to the Editor' column of *The American Statistician*. Selvin formulated the problem in terms of 'boxes', and this is what we copied when we stated it as a problem, but the problem became famous in its '*car and goats behind three doors*' formulation, which we have felt obliged to give.

Incidentally, in his 'letter', Selvin misspelt Monty Hall's name, calling him 'Monte Hall': I consider this a feather in Selvin's hat – statisticians (and mathematicians) are unlikely to watch game shows.

The Monty Hall formulation of MHP is by far the most famous, but the problem had appeared in several guises before Selvin posed it in his letter: in particular, Martin Gardner published it in his 'Mathematical Games' column in 1959 as *The Three Prisoners Problem*. As I have already mentioned, the fame of MHP is due to Marilyn vos Savant, presented by the *Guinness Book of World Records* as the person with the highest IQ ever recorded, who gained an awful lot of publicity from the stupidity and high-handedness of some mathematicians. In addition, many (far too many!) academics jumped on the bandwagon and wrote papers about the problem and its further variants. These publications even include a book!

The claimed (but non-existent) difficulty of the problem is often justified by quoting from a 1999 article by Andrew Vázsonyi, who was a friend of Paul Erdős for sixty years, in which Vázsonyi describes that it took Erdős ages to understand that switching indeed increases the chances of winning the car behind the door. In fact, according to Vázsonyi, Erdős could not understand the explanation based on decision trees. I do not find this that surprising, as I have no doubt that Erdős had never heard of a decision tree, no matter how trivial that concept is. However, I know that he would have understood the solution presented above in a jiffy.

In fact, a 'decision tree' solution of this puzzle can also be described easily: the formality seems to make the solution mindless. Let A, B and C be the three doors. As they are indistinguishable, we may assume that the car is behind door A. [Equivalently, we may name the door hiding the car A.] Then the contestant has three possibilities: he may choose door A or door B or door C. Again, the doors are indistinguishable, so each choice has the same probability, $1/3$. In each of these cases, the contestant may *stay with his choice* or *switch* to the *unique* third door. For example, choosing door B and *not switching* gives us the sequence B, NS, resulting in a goat: $f(B, NS) = G$, and *switching* results in the car: $f(B, S) = C$. Here are all the possibilities: $f(A, NS) = C$, $f(A, S) = G$, $f(B, NS) = G$, $f(B, S) = C$, $f(C, NS) = G$, $f(C, S) = C$. Thus, if the contestant switches, he wins a car if he first chooses doors B or C, which has probability $2/3$, and if he stays with his choice, he wins the car if he chooses A, an event of probability $1/3$.

Thus, the MHP was a resounding success for Marilyn vos Savant, the 'world's smartest woman': her website gives plenty of evidence of that. Sadly, she badly blotted her copybook when she went on to write a book about Andrew Wiles's proof of Fermat's Last Theorem (FLT). Let me quote a few lines from the review of this book by Nigel Boston and Andrew Granville, which appeared in *American Mathematical Monthly* volume 102, pages 470–473, in 1995.

> "[Marilyn vos Savant] has challenged the orthodoxy of the mathematics world by refuting Wiles's purported proof of FLT, by claiming that it is wrong because it is illogical, relying, she believes, on ridiculous inconsistencies accepted by mathematicians. For example, the concept of a non-Euclidean geometry. For another example, proofs using induction. [?!] . . ."

> "Thus she concluded that Wiles gave a 'hyperbolic method of proving FLT.' In fact, her central theme is that non-Euclidean geometry, and indeed any mathematics related to non-Euclidean geometry, is nonsense. Her thesis seems to be that, since it was proved in 1882 that 'squaring the circle' is impossible in a Euclidean setting, and since Bolyai managed to 'square the circle' in an appropriate non-Euclidean geometry, thus non-Euclidean geometry is inconsistent with Euclidean geometry. However since FLT is a statement consistent with regular geometry, it cannot be proved by arguments that involve any non-Euclidean geometry. After all, 'one of the founders of hyperbolic geometry [J. Bolyai] managed to square the circle?! Then why is it known as such a famous impossibility?' Therefore, she concludes, 'if we reject a hyperbolic method of squaring the circle, we should also reject a hyperbolic proof of FLT!'. This is typical of the inane reasoning (and hyperbole) that pervades this book."

Many people will realize that the statements of MvS mentioned above are utter nonsense. I hope that this cautionary tale gives a little indication that the jump from the Monty Hall Problem to Andrew Wiles's proof of Fermat's Last Theorem is beyond imagination!

References

Bailey, H., Monty Hall uses a mixed strategy *Math. Mag.* **73** (2000) 135–141.

Bar-Hillel M. and R. Falk, Some teasers concerning conditional probabilities, *Cognition* **11** (1982) 109–122.

Burns, B. and M. Wieth, The collider principle in causal reasoning: Why the Monty Hall Problem is so hard, *J. Experi. Psychol., Gen.* **103** (2004) 436–449.

Gardner, M., Problems involving questions of probability and ambiguity, *Scientific American* **201** (April 1959) 174–182.

Gardner, M., Mathematical games, *Scientific American* **201** (October 1959) 180–182.

Gardner, M., Mathematical games, *Scientific American* (November 1959) 188.

Lucas, S., J. Rosenhouse and A. Schepler, The Monty Hall problem, reconsidered, *Math. Mag.* **82** (2009) 332–342.

Rosenhouse, J., *The Monty Hall Problem. The Remarkable Story of Math's Most Contentious Brainteaser*, Oxford University Press (2009).

Selvin, S., Letters to the Editor, *The American Statistician* **29** (1975), 67.

Selvin, S., *Survival Analysis for Epidemiologic and Medical Research. A Practical Guide*, Practical Guides to Biostatistics and Epidemiology, Cambridge University Press (2008).

vos Savant, M., *The World's Most Famous Math Problem (The Proof of Fermat's Last Theorem and Other Mathematical Mysteries)*, St. Martin's Press (1993).

vos Savant, M., *The Power of Logical Thinking*, St. Martin's Press (1996).

60. Divisibility in a Sequence of Integers

Let $a_1 < a_2 < \cdots$ be an infinite sequence of natural numbers. Then there exists either an infinite subsequence in which no integer divides another, or an infinite subsequence in which every term divides all subsequent terms.

Proof. Consider all a_i that divide no other integer in the sequence. If there are infinitely many such a_i, we have our subsequence in which no integer divides another. Otherwise there are finitely many such a_i: in that case delete them from the sequence together with all their divisors. This leaves infinitely many terms. Pick one of them, a_{i_1}, say. Having picked $a_{i_1}, \ldots, a_{i_k}$, with each term (except a_{i_k}) dividing the next, choose for $a_{i_{k+1}}$ a term that is a multiple of a_{i_k}. This gives us an infinite subsequence in which every term divides all subsequent terms. $\qquad\square$

Notes. This problem was set in the *American Mathematical Monthly* by Paul Erdős in 1949. Three solutions were published: the first, the solution we have presented, by R.S. Lehman, the second by G.A. Hedlund and the third by R.C. Buck. To today's young mathematicians this exercise is just too trivial: it is an immediate consequence of Ramsey's theorem for infinite sets. Indeed, consider the complete graph on the sequence (a_i), with an edge $a_i a_j$ coloured *red* if a_i divides a_j or a_j divides a_i, and *blue* otherwise. By Ramsey's theorem there is a monochromatic infinite subgraph: this is exactly what we had to find. It is surprising that Erdős posed a problem that was an immediate consequence of Ramsey's theorem since, with Szekeres, he discovered this result independently of Ramsey.

Nevertheless, Lehman's proof above is considerably simpler. The third proof establishes a formally stronger result: in fact, it is just Lehman's proof in a more general setting. Here is the 'more general' result. *Let $(X, <)$ be an infinite poset in which every element dominates only finitely many other elements. Then X* *contains an infinite chain or an infinite antichain.* [In a chain no two elements a and b are incomparable, i.e. either $a < b$ or $b < a$, and in an antichain any two elements are incomparable.]

Setting this problem to undergraduates doing the Graph Theory course at Cambridge, most undergraduates simply applied Ramsey's theorem, but some (rather few) gave me much pleasure by noticing that much less suffices to solve the problem.

References

Erdős, P. and G. Szekeres, A combinatorial problem in geometry, *Compositio Math.* **2** (1935) 463–470.

Erdős, P., R.S. Lehman, G.A. Hedlund and R.C. Buck, Advanced Problems and Solutions: Solutions: 4330, *Amer. Math. Monthly* **57** (1950) 493–494.

Ramsey, F.P., On a problem of formal logic, *Proc. London Math. Soc. (2)* **30** (1929) 264–286.

61. Moving Sofa Problem

A long passage of unit width has a right-angled bend in it. A flat rigid plate P (made up of one piece) of area A can be manœuvered from one end of the passage to the other. Then

$$A < 2\sqrt{2} \sim 2.8284.$$

Also, there is a suitable plate of area at least

$$\frac{\pi}{2} + \frac{2}{\pi} \sim 2.2074.$$

Proof. (i) Let us start with a proof of the first inequality, that the area of the plate is strictly less than $2\sqrt{2}$. Before our plate P reaches the corner, it is contained in an infinite strip S_b of width at most 1 determined by two lines parallel to the walls of the (first part of the) corridor. Continuing on after the corner, there is a similar strip S_a. We consider these strips (which may be the same but need not be) attached to P: as the plate is moved and turned, so do the strips. At some stage of the move, the angle α_b of the first strip S_b formed with the walls of the first part of the corridor is equal to the angle α_a the second strip S_a formed with the walls of the second corridor. At this stage our plate is contained in the intersection of the two strips (which is either a parallelogram or an infinite strip) and the corridor. Since P is made up of one piece, this intersection has to be connected, as in Figure 36a. Clearly, the area of this (connected) intersection is at most as large as the intersection in the case when $\alpha_a = \alpha_b = \pi/4$ (so the strips S_a and S_b coincide), and the inner corner of the corridor is on the 'outer' boundary of the strip, as in Figure 36b. We have strict inequality since in this case the intersection falls into two separate pieces. Finally, the area of each of the two rhombi making up the intersection has area $\sqrt{2}$ since it has height 1 and side $\sqrt{2}$. Thus $A < 2\sqrt{2}$.

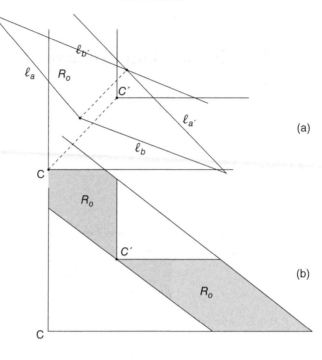

(a)

(b)

Figure 36

(ii) Consider the plate in Figure 37, resembling an old-fashioned telephone handset. It is inscribed in a rectangle of height 1 and width $2 + 4/\pi$: it consists of two quarter circles of unit radius (on the two sides), and the middle is a rectangle of height 1 and width $4/\pi$, from which we cut out a half-circle of radius $2/\pi$. This plate, which is often called the 'Hammersley sofa', has area $\pi/4 + \left(4/\pi - (\pi/2)(2/\pi)^2\right) + \pi/4 = \pi/2 + 2/\pi \sim 2.2074$. For an animated illustration of the way Hammersley's sofa is moved around the bend, see https://en.wikipedia.org/wiki/Moving-sofa-problem.   □

Notes. This problem has been popular among professional and amateur mathematicians for about half a century, and has an extensive literature. Its origin is not clear-cut, it is likely that it arose independently in several places in the early 1960s, perhaps even earlier. It seems that its first appearance in print was in 1966, when Leo Moser posed it as a problem in *SIAM Review*, but it was known in Cambridge already in 1964 (and appeared in Hallard Croft's mimeographed notes in 1967, where he wrote about 'Shephard's Piano', an improvement on 'Hammersley's Sofa' above), and was also mentioned by John Hammersley in

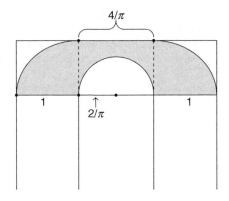

Figure 37 Hammersley sofa, of area $\pi/2 + 2/\pi$.

his famous lecture to *The Institute of Mathematics and its Applications* in June 1967. As a tribute to Hammersley, our first formulation is almost verbatim from his 1968 article, which was an expanded version of his lecture.

Although privately Hammersley conjectured that his sofa was optimal, his construction was soon beaten by Shephard's piano. An important development came in 1992, when Gerver gave a necessary condition for a plate to have the greatest possible area if it can be moved around a right-angled corner in a hallway of unit width. Furthermore, Gerver constructed a plate of area 2.2195 . . . , bounded by 18 analytic curves, which satisfies this condition. He also proved that there is a plate of maximal area (an unpublished result of John Conway and Richard Guy) and conjectured that his plate had maximal area.

More recently, Romik, and Kallus and Romik proved substantial results about the sofa problem. In particular, Kallus and Romik used a computer-assisted method to improve Hammersley's upper bound out of recognition to 2.37. Their method can also be used to rigorously prove further upper bounds that converge to the correct value.

References

Croft, H. T., K.J. Falconer and R.K. Guy, *Unsolved Problems in Geometry*, corrected reprint of the 1991 original, *Problem Books in Mathematics, Unsolved Problems in Intuitive Mathematics*, II. Springer-Verlag (1994).

Gerver, J.L., On moving a sofa around a corner, *Geom. Dedicata* **42** (1992) 267–283.

Hammersley, J.M., On the enfeeblement of mathematical skills by 'Modern Mathematics' and by similar soft intellectual trash in schools and universities, *Bull. Inst. Math. Appl.* **4** (1968) 66–85.

Kallus, Y. and D. Romik, Improved upper bounds in the moving sofa problem, *Adv. Math.* **340** (2018) 960–982.

Moser, L., Moving furniture through a hallway, *SIAM Rev.* **8** (1966) 381–381.

Romik, D., Differential equations and exact solutions in the moving sofa problem, *Exp. Math.* **27** (2018) 316–330.

Sebastian, J.D., Moving furniture through a hallway (Leo Moser), *SIAM Rev.* **12** (1970) 582–586.

62. Minimum Least Common Multiple

Let $a_1 < a_2 < \cdots < a_n \leq 2n$ be a sequence of $n \geq 5$ positive integers. (i) Then there are integers $a_i < a_j$ such that

$$[a_i, a_j] \leq 6(\lfloor n/2 \rfloor + 1).$$

(ii) *This inequality is best possible.*

Proof. (i) Note that if $a_1 = n + 1$, i.e. $n + 1, n + 2, \ldots, 2n$ are our numbers, then both $2(\lfloor n/2 \rfloor + 1)$ and $3(\lfloor n/2 \rfloor + 1)$ appear in our sequence, so

$$\min_{i<j}[a_i, a_j] \leq 6(\lfloor n/2 \rfloor + 1),$$

showing that our inequality holds.

Hence, we may suppose that $a_1 \leq n$. For each i, let k_i be such that $n + 1 \leq k_i a_i \leq 2n$. If some two of these numbers coincide, i.e. $m = k_i a_i = k_j a_j$ for some $1 \leq i < j \leq n$ with $n + 1 \leq m \leq 2n$, then $[a_i, a_j] \leq m \leq 2n$ — an overkill! Otherwise, $k_1 a_1, \ldots, k_n a_n$ are n numbers between $n + 1$ and $2n$, so $k_1 a_1, \ldots, k_n a_n$ is an enumeration of $n + 1, n + 2, \ldots, 2n$. In particular, there are indices i and j such that

$$k_i a_i = 2(\lfloor n/2 \rfloor + 1) \qquad \text{and} \qquad k_j a_j = 3(\lfloor n/2 \rfloor + 1),$$

implying that $[a_i, a_j] \leq 6(\lfloor n/2 \rfloor + 1)$.

(ii) To show that the inequality is best possible, let us return to the sequence $n + 1, n + 2, \ldots, 2n$. We claim that for this sequence the least common multiple of two terms is at least $6(\lfloor n/2 \rfloor + 1)$. Suppose for a contradiction that this is not the case: there are terms $n + i < n + j$ with $m = [n + i, n + j] < 6(\lfloor n/2 \rfloor + 1)$. Set $m = k_i(n + i) = k_j(n + j)$, so that $2 \leq k_j < k_i$.

Since $[n + 1, n + 2] = (n + 1)(n + 2) > 6(\lfloor n/2 \rfloor + 1)$, we have $j \geq 3$. Also,

$$4(n + 1) > 6(\lfloor n/2 \rfloor + 1),$$

191

so $k_i = 3$ and $k_j = 2$, and so $m = 3(n + i) = 2(n + j)$. This tells us that $n + i$ is even. Since $m = 3(n + i) < 6(\lfloor n/2 \rfloor + 1)$, we must have $n + i < 2(\lfloor n/2 \rfloor + 1)$ and so $n + i \le 2\lfloor n/2 \rfloor \le n$. This contradiction completes our proof. $\square$

Notes. This result, a companion of the one in Problem 56, is again due to Paul Erdős. There we demanded that every lowest common multiple was large, here we claim that there is one that cannot be too large. When Erdős set it in 1937, he made a slight mistake: this was corrected by Phelps in his solution published in 1958, and reproduced above.

Reference

Erdős, P. and C.R. Phelps, Advanced Problems and Solutions: Solutions: 3834, *Amer. Math. Monthly* **65** (1958) 47–48.

63. Vieta Jumping

Let a and b be positive integers such that $q = (a^2 + b^2)/(ab + 1)$ is also an integer. Then q is a perfect square.

Proof. Suppose for a contradiction that q is not a square. Let (a, b) be a pair of positive integers satisfying

$$q = \frac{a^2 + b^2}{ab + 1} \tag{1}$$

such that their sum $s = a + b$ is minimal.

Note that if $a = b$ then $q = 2a^2/(a^2 + 1) < 2$, so $q = 1$, contradicting our assumption that q is not a square. Therefore we may assume that $1 \le a < b$. Although we shall not need it, note that this implies that $a > 1$ since if we had $a = 1 < b$ then $q = (b^2 + 1)/(b + 1)$ would not be an integer.

Replacing b by a variable x in (1), we get a quadratic equation

$$x^2 - aqx + (a^2 - q) = 0 \tag{2}$$

with two (possibly equal) roots x_1 and x_2 satisfying $x_1 + x_2 = aq$ and $x_1 x_2 = a^2 - q$. We know that b is a root, say, $x_1 = b$, so the other root has to be $x_2 = aq - b$, which is also an integer. Our aim is to show that x_2 is a positive integer less than b: in this case we are done, since choosing x_2 for b, the new pair satisfies (1) and their sum is smaller than s.

Equation (2) tells us that $x_2 = (a^2 - q)/b$, which is not zero, since q is not a perfect square. Also,

$$q = \frac{x_2^2 + a^2}{ax_2 + 1}$$

tells us that $x_2 \ge 1$ since $x_2 \le -1$ would make q negative. Finally,

$$x_2 = \frac{a^2 - q}{b} < \frac{a^2}{b} < a < b,$$

which is a contradiction with plenty to spare. This completes our proof.   □

Notes. This is the infamous Problem 6 at the 1988 IMO (International Mathematical Olympiad). Without a hint, this was an unreasonably difficult problem to set; nevertheless, eleven of the 268 competitors (aged at most eighteen) solved it perfectly within the hour and a half hour allotted for the problem.

The name of this exercise, 'Vieta Jumping', refers to the fact that the proof was based on jumping from one root of a quadratic equation to another. The equations connecting the roots of polynomials to the polynomial coefficients are usually attributed to the French lawyer and mathematician François Viète (1540–1603) whom it is usual to call by his Latin name, Franciscus Vieta. We obtain the Vieta formulae immediately if we expand the factorization $(x - x_1)(x - x_2) \cdots (x - x_n)$ of a polynomial f, where $x_1, \ldots x_n$ are the roots. Thus the two roots x_1 and x_2 of $x^2 + bx + c$ satisfy $x_1 + x_2 = -b$ and $x_1 x_2 = c$, and the three roots x_1, x_2, x_3 of the cubic polynomial $x^3 + bx^2 + cx + d$ satisfy $x_1 + x_2 + x_3 = -b$, $x_1 x_2 + x_2 x_3 + x_3 x_1 = c$ and $x_1 x_2 x_3 = -d$.

64. Infinite Primitive Sequences

Let $a_1 < a_2 < \cdots$ be an infinite sequence of natural numbers such that no term divides another, i.e. let $A = \{a_1, a_2, \ldots\}$ be a primitive sequence of natural numbers. Then

$$\sum_{i=1}^{\infty} \frac{1}{a_i} \prod_{p \leq p_i} \left(1 - \frac{1}{p}\right) \leq 1, \tag{1}$$

where p_i is the largest prime divisor of a_i and the product is over all primes p less than p_i.

Proof. Suppose that (1) does not hold. Then there is an integer ℓ such that

$$\sum_{i=1}^{\ell} \frac{1}{a_i} \prod_{p \leq p_i} \left(1 - \frac{1}{p}\right) > 1. \tag{2}$$

Set $p_{\max} = \max_{i \leq \ell} p_i$,

$$n = \left(\prod_{i=1}^{\ell} a_i\right)\left(\prod_{p \leq p_{\max}} p\right)$$

and for $1 \leq i \leq \ell$ define

$$M_i = \{ta_i \leq n : \text{ every prime factor of } t \text{ is greater than } p_i\}.$$

Then $M_1, M_2, \ldots, M_\ell$ are pairwise disjoint subsets on $[n]$. Indeed, suppose $sa_i = ta_j$ with $sa_i \in M_i$, $ta_j \in M_j$, where $i \neq j$ and $p_i \leq p_j$. Then a_i and t are relatively prime and $a_i | ta_j$. Hence $a_i | a_j$ – a contradiction. Finally, since

$$\prod_{p \leq p_i} p \text{ divides } n/a_i,$$

we have

$$|M_i| = \frac{n}{a_i} \prod_{p \leq p_i} \left(1 - \frac{1}{p}\right),$$

195

and so

$$n \geq \left| \bigcup_{i=1}^{\ell} M_i \right| = \sum_{i=1}^{m} \frac{n}{a_i} \prod_{p \leq p_i} \left(1 - \frac{1}{p} \right).$$

This inequality contradicts (2) and so completes our proof.   □

Notes. The result is from one of the early papers of Paul Erdős. In fact, the theorem he stated is that if in the sequence $a_1 < a_2 < \cdots$ no term divides another then

$$\sum_n \frac{1}{a_n \log a_n} \leq c$$

for some constant c. This is immediate from the inequality above and the fact that

$$\prod_{p \leq p_n} \left(1 - \frac{1}{p} \right) \geq \prod_{p \leq a_n} \left(1 - \frac{1}{p} \right) \geq \frac{c}{\log a_n}.$$

In 1988 Erdős suggested that the optimal primitive sequence is the sequence of primes in the sense that if $1 < a_1 < a_2 < \cdots$ is a primitive sequence and $p_1 = 2, p_2 = 3, p_3 = 5, \ldots$ is the sequence of primes then

$$\sum_n \frac{1}{a_n \log a_n} \leq \sum_n \frac{1}{p_n \log p_n} = 1.636616 \cdots.$$

This notorious 'Erdős conjecture' is still unsolved, although Erdős and Zhang, Clark, and, most importantly, Lichtman and Pomerance made much progress towards a proof. In particular, Lichtman and Pomerance proved that every primitive sequence (a_n) satisfies

$$\sum_n \frac{1}{a_n \log a_n} < e^{\gamma} = 1.781072 \cdots,$$

where γ is Euler's constant.

References

Clark, D.A., An upper bound of $\sum 1/(a_i \log a_i)$ for primitive sequences, *Proc. Amer. Math. Soc.* **123** (1995) 363–365.

Erdős, P., Note on sequences of integers no one of which is divisible by any other, *J. London Math. Soc.* **10** (1935) 126–128.

Erdős, P. and Z. Zhang, Upper bound of $\sum 1/(a_i \log a_i)$ for primitive sequences, *Proc. Amer. Math. Soc.* **117** (1993) 891–895.

Lichtman, J.D. and C. Pomerance, The Erdős conjecture for primitive sets, *Proc. Amer. Math. Soc. Ser. B* **6** (2019) 1–14.

65. Primitive Sequences with a Small Term

Let $1 < a_1 < a_2 < \cdots < a_n \leq 2n$ be a primitive sequence, i.e. such that no term divides another and the sequence has maximal length in the interval $[1, 2n]$. Then $a_1 \geq 2^k$, where k is defined by the inequalities $3^k < 2n < 3^{k+1}$.

Proof. Write $a_i = 2^{\alpha_i} b_i$, where α_i is a non-negative integer and b_i is odd. Then, to avoid divisibility, the factors b_i have to be distinct, so

$$a_i = 2^{\alpha_i} b_i, \qquad \text{where} \quad b_i = 1, 3, \ldots, 2n - 1 \quad \text{in some order.}$$

First, consider the a_i with $b_i = 1, 3, 3^2, \ldots, 3^k$. These terms can be written as

$$2^{\alpha_i} 3^i, \qquad i = 0, 1, \ldots, k.$$

To avoid divisibility, we must have $\alpha_0 > \alpha_1 > \cdots > \alpha_k$. Therefore $\alpha_i \geq k - i$ and so

$$2^{\alpha_i} 3^i \geq 2^{k-i} 3^i \geq 2^k.$$

Thus these terms are at least as large as claimed.

Can one of the other terms be smaller than 2^k? Suppose that this is the case, say

$$a = 2^{\alpha} b < 2^k,$$

where b is an odd integer which is not a power of 3, so is at least 5. Here we have deliberately omitted the suffices of a, b and α, as they will play a pivotal role in the argument that follows. Note that

$$5 \leq b < 2^{k-\alpha}, \qquad \text{so} \quad k - \alpha \geq 3.$$

The b_i, the odd factors, are *all* the distinct odd integers less than $2n$, so the numbers

$$b, \ 3b, \ 3^2 b, \ \ldots, \ 3^{\alpha+1} b$$

197

all appear as odd factors, since even the largest is less than $2n$, as shown by the following sequence of inequalities:

$$3^{\alpha+1}b < 3^{\alpha+1}2^{k-\alpha} < 3^{\alpha+1}3^2 2^{k-\alpha-3} \le 3^k < 2n.$$

Consequently, these $\alpha + 2$ odd factors have to give the following subset of the original sequence:

$$b\, 2^{\gamma_1}, \quad 3b\, 2^{\gamma_2}, \quad 3^2 b\, 2^{\gamma_3}, \quad \ldots, \quad 3^{\alpha+1}b\, 2^{\gamma_{\alpha+2}}.$$

We know two properties of this sequence, which lead to a contradiction. First, our starting point was that $a = 2^\alpha b$ is in our sequence, so we must have $\gamma_1 = \alpha$. Second, to avoid divisibility, the γ_i must be strictly decreasing, i.e. we must have

$$\alpha = \gamma_1 > \gamma_2 > \cdots > \gamma_{\alpha+2} \ge 0,$$

which is impossible.   □

Notes. The problem above was set in *The Monthly* in 1939, and the journal published a solution by Emma Lehmer, who was in Cambridge at the time with her husband. In the problem Erdős also asked for a proof that this bound was best possible. Somewhat surprisingly, this is not that easy to see. Here is Lehmer's construction as a two-dimensional array:

$$a_{ij} = 2^{k_i - j}3^j \omega_i, \qquad \text{where} \quad \begin{cases} \omega_i < 2n \text{ and prime to } 6; \\ 3^{k_i} < \frac{2n}{\omega_i} < 3^{k_i+1}; \\ j = 0, 1, \ldots, k_i. \end{cases}$$

As Lehmer says, 'It can be easily verified that this set satisfies the conditions of the problem and has 2^k for its least element. For example, for $n = 15$ we have as our set the numbers 8, 10, 11, 12, 13, 14, 15, 17, 18, 19, 21, 23, 25, 27, 29'.

Reference

Erdős, P. and E. Lehmer, Solution of Problem 3820, *Amer. Math. Monthly* **46** (1939) 240–241.

66. Hypertrees

Let G be an r-uniform hypergraph containing no copy of a certain r-tree T with m edges. Show that G is k-colourable for k = 2(r − 1)(m − 1) + 1.

Proof. Apply induction on m. For $m = 1$ the assertion is trivial, since G contains no edges, so let us turn to the induction step. Suppose that $m \geq 2$, and let T have edges $E_1, \ldots, E_m$, as in the statement of our problem. Denote by T' the subtree of T formed by the edges $E_1, \ldots, E_{m-1}$ (and so having vertex set $\bigcup_{i=1}^{m-1} E_i$), and write x_0 for the vertex common to E_m and $\bigcup_{i=1}^{m-1} E_i$. Let $W \subset V(G)$ be the set of vertices y_0 of G such that G contains a copy T_{y_0} of T', with y_0 corresponding to x_0.

Define a graph (not hypergraph!) H on $V(G)$ by joining y_0 to all $(r-1)(m-1)$ vertices of T_{y_0}, $y_0 \in W$. Set $U = V(G) \backslash W$ and $G' = G[U]$. Note that G' contains no copy of T', so by the induction hypothesis it is a k'-colourable hypergraph, where $k' = 2(r - 1)(m - 2) + 1$. Also, by the construction of H, we may write W as $W = \{y_1, \ldots, y_\ell\}$, with y_i sending at most $2(r - 1)(m - 1)$ edges to $U \cup \{y_1, \ldots, y_{i-1}\}$ for every i, $1 \leq i \leq \ell$. But then the k'-colouring of the hypergraph G' can be extended to a k-colouring of the graph H, where $k = 2(r - 1)(m - 1) + 1$. This k-colouring is, in fact, a proper colouring of the hypergraph G, since every (hyper)edge of G containing a vertex y_0 of W has to contain an edge of H. Indeed, otherwise G would contain T_{y_0}, together with an edge meeting the vertices of T_{y_0} in precisely y_0, and so G would contain a copy of T. □

67. Subtrees

(i) *A tree on n vertices contains at least $\binom{n+1}{2}$ subtrees (each with at least one vertex). For every $n \geq 1$, the path on n vertices is the unique extremal tree.*

(ii) *A tree on n vertices contains at most $2^{n-1} + n - 1$ subtrees (each with at least one vertex). For every $n \geq 1$, the star is the unique tree with the maximal number of subtrees.*

Proof. (i) Let T_n be a tree with vertex set $[n] = \{1, \ldots, n\}$. For all $1 \leq i \leq j \leq n$ there is a path in T_n from i to j.

Figure 38 A path and a star, each on 7 vertices.

In a path of order n, these are the only subtrees; in every other tree, there are also subtrees with three end vertices.

(ii) There are n subtrees with one vertex each. Every other subtree is determined by the non-empty set of edges it contains: as there are $n - 1$ edges, there are $2^{n-1} - 1$ non-empty sets of edges and at most $n + 2^{n-1} - 1$ subtrees, as claimed.

The star on n vertices has $n + 2^{n-1} - 1$ subtrees. Conversely, a tree T of order n with $n + 2^{n-1} - 1$ subtrees is such that every pair of edges forms a subtree, i.e. every two edges of T share a vertex. But then T is the star on n vertices (Figure 38). $\qquad\square$

68. All in a Row

The maximum probability that every *student guesses correctly is* 1/2.

Proof. As the guess of the first student is made before he is given any information (he calls out first!), the probability that he guesses correctly, whatever he says, is 1/2. So the probability that *every* student guesses correctly is at most 1/2. Furthermore, this 1/2 probability can be achieved only if every other student, after the first, guesses correctly *with probability* 1.

Thus, our task is to show that with his guess the first student can start an avalanche in which every (other) student guesses correctly. So what information can the first student give to his class mates with a single guess 'white' or 'black'? He can tell them the *parity* of the number of white hats he sees: say, 'white' means 'I see an even number of white hats'. This information enables the second to deduce the colour of her hat: if she sees an odd number of white hats, her hat must be white, so she calls out 'white', otherwise 'black'.

After the first two calls the third knows the parity of the number of hats on the last eighteen students: if this parity is the same as the parity of the number of hats on the last seventeen, i.e. on the students she sees, she calls out 'black', otherwise her guess is 'white'.

Formally, if a student (other than the first) hears 'white' called out c times and sees s white hats, then his hat is white if $c + s$ is even, and black if $c + s$ is odd.   □

Notes. Pity the first student: he had to guess the colour of his own hat based on what he saw on his class mates, so he had no chance of increasing the probability of success from $1/2$. Conversely, he had no chance of *decreasing* the probability from $1/2$ either, so he was free to use his guess only for the benefit of the others – this is exactly what he did in the strategy above.

69. An American Story

The twenty Death Row prisoners have a strategy which gives them a better than even chance of escaping execution.

Solution. Yes, there is a suitable strategy. In particular the following is one such. They assign the boxes to themselves at random, so that every one of them has his 'own' box. Needless to say, this one-to-one assignment has nothing to do with the names in the boxes. Every prisoner first opens his 'own' box. If he finds his name, he may as well stop. Otherwise, next he opens the box assigned to the prisoner whose name he finds there, and so on. To spell this out, writing $P_1, \ldots, P_{20}$ for the names of the prisoners, and $B_1, \ldots, B_{20}$ for the boxes, if B_5 contains P_8, B_8 contains P_3, B_3 contains P_{19}, and B_{19} contains P_5, then prisoner P_5 first opens B_5, then B_8, B_3 and B_{19}, upon which he gives a sigh of relief, as he has just found his own name. With this, P_5 may as well stop, or else open another eight boxes out of curiosity.

All that remains is to check that this strategy succeeds with probability at least $1/2$. To this end, note that the random assignment of boxes to prisoners is equivalent to a permutation of $B_1, \ldots, B_{20}$: to obtain this permutation, map B_i into B_j if B_i contains P_j. Then the strategy of the prisoners is to open boxes along cycles: prisoner P_i finds his name if the cycle containing B_i has length at most 12. Hence, the prisoners gain their reprieve if the permutation they take at random contains no *long cycle*, i.e. no cycle of length at least 13.

It is easy to calculate the probability p_L that our random permutation of length 20 contains a long cycle. Indeed, as a permutation contains at most one long cycle, p_L is just the expected number of long cycles. Now, the expected number of cycles of length $\ell \geq 2$ in a random permutation of $[n] = \{1, \ldots, n\}$ is

$$\binom{n}{\ell} \frac{\ell!}{\ell} (n - \ell)! / n! = \frac{1}{\ell},$$

203

since there are $\ell!/\ell$ *oriented cycles* on ℓ given vertices. Consequently,

$$p_L = \frac{1}{13} + \frac{1}{14} + \cdots + \frac{1}{20} < 0.5,$$

as claimed.   □

Notes. This puzzle was first considered by Peter Bro Miltersen in connection with some open problems that he and Anna Gál were working on, and appeared in a joint paper by Gál and Miltersen. The unexpected solution above was given by Sven Skyum. I first heard about this problem from Gál during a meeting in Oberwolfach. For more about some related problems, see the paper of Goyal and Saks.

References

Gál, A. and P.B. Miltersen, The cell probe complexity of succinct data structures. In *Proc. Int. Coll. Automata, Languages and Programming (ICALP)*, 332–344, Lecture Notes in Comput. Sci. **2719**, Springer (2003).

Gál, A. and P.B. Miltersen, The cell probe complexity of succinct data structures, *Theoret. Comput. Sci.* **379** (2007) 405–417.

Goyal, N. and M. Saks, A parallel search game, *Random Struct. Algorithms* **27** (2005) 227–234.

70. Six Equal Parts

Let S be a set of 6k points in general position in the plane. Then there are three concurrent lines that partition S into six equal parts.

Proof. This result is an easy consequence of an analogous assertion about finite measures. As we wish to use as little 'sophisticated' mathematics as possible, we shall state this result for very special measures indeed. Let f be a continuous strictly positive function on the plane whose integral over $\mathbb{R}^2$ is 1. (For example, $f(x,y)$ could be the density function of two independent standard normal random variables, so that $f(x,y) = \frac{1}{2\pi}e^{-(x^2+y^2)/2}$.) Define the *measure* $\mu(U)$ of an open set $U \subset \mathbb{R}^2$ as $\mu(U) = \int_U f(x,y)\mathrm{d}x\mathrm{d}y$. In fact, for this discussion it suffices to define $\mu(U)$ only in the case when U is the intersection of finitely many (open or closed) half-planes.

Claim *Let μ be a probability measure on $\mathbb{R}^2$ as above. Then there are three concurrent lines that partition the plane into six segments of equal measure.*

To prove this, all we shall use about μ is that the measure of a sector of the plane (the set of points between two rays emanating from the same point) is a continuous function (in the obvious sense) of the sector.

For every unit vector u there is a unique line ℓ_u in the direction of u which cuts the plane into two equal parts (so each part has measure $1/2$). For every point $x \in \ell_u$, there are six unique rays (half-lines) from x, say $r_1(u,x),\dots,r_6(u,x)$, which cut the plane into six sectors of equal measure. Write $w_i(u,x)$ for the direction of ray $r_i(u,x)$, so that $w_1(u,x) = u$. Note that $w_4(u,x) = -u$: the line ℓ_u is the union of the two rays $r_1(u,x)$ and $r_4(u,x)$.

Our next aim is to find another line from the next two rays. We claim that there is a unique point $x_u \in \ell_u$ such that $w_5(u,x_u) = -w_2(u,x_u)$, so that the second and fourth rays are collinear. Indeed, if x is moved along ℓ_u in the u direction then the angle formed by $w_1(u,x)$ and $w_2(u,x)$ is continuously and

205

strictly increasing from 0 to π, and the angle formed by $w_4(u, x) = -w_1(u, x)$ and $w_5(u, x)$ is continuously and strictly decreasing from π to 0, so at a unique point x_u the two angles are indeed equal.

So far, we have produced two of our three lines: it remains to find a direction u_0 at which $w_3(u_0, x_{u_0})$ and $w_6(u_0, x_{u_0})$ also point in opposite directions. Once again, continuity and 'swapping places' save the day. As u is rotated along the unit circle, the line ℓ_u and the point x_u on this line move continuously, and so do the directions $w_i(u, x_u)$. When the starting position u becomes $-u$, the unit vectors $w_3(u, x_u)$ and $w_6(u, x_u)$ are the other way round, so there is a direction u_0 during the move such that $w_6(u_0, x_{u_0}) = -w_3(u_0, x_{u_0})$. For this direction u_0 the six rays $r_1(u_0, x_{u_0}), \ldots, r_6(u_0, x_{u_0})$ emanating from x_{u_0} form three lines that partition the plane into six parts of equal measure. This completes the proof of our Claim.

Armed with it, our assertion concerning the set S of $6k$ points in general position is easily proved. Indeed, replace each point by a disc of radius $r > 0$, where r is small enough to guarantee that no line meets three of these discs. Give each disc measure $(1 - \varepsilon)/6k$, where $\varepsilon < 1/100k$, say, and the rest of the plane measure ε: this is easily done in any number of ways. Then the three lines guaranteed by the Claim above partition our set S of $6k$ points into six equal parts. □

Notes. This is a slight extension of a result of Buck and Buck, who proved the analogous six-partition for convex sets. The measure-theoretic version above was noticed by Ceder; in the proof we followed Bukh.

Buck and Buck also proved that no three lines partition a bounded convex set in the plane into *seven* parts of equal measure. Needless to say, the three lines in question are not assumed to be concurrent, but bound a triangle, as in Figure 39.

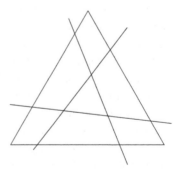

Figure 39 Three lines that partition a triangle into seven roughly equal parts.

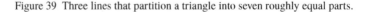

References

Bárány, I., A generalization of Carathéodory's theorem, *Discrete Math.* **40** (1982) 141–152.

Boros, E. and Z. Füredi, The number of triangles covering the center of an *n*-set, *Geom. Dedicata* **17** (1984) 69–77.

Buck, R.C. and E.F. Buck, Equipartition of convex sets, *Math. Mag.* (1949) 195–198.

Bukh, B., A point in many triangles, *Electronic J. Comb.* **13** (2006) #10, 3 pp.

Ceder, J.G., Generalized sixpartite problems, *Bol. Soc. Mat. Mexicana (2)* **9** (1964) 28–32.

71. Products of Real Polynomials

(i) *The degree of the product of two real polynomials in several variables is the sum of the degrees.*

(ii) *The degree of a sum of squares of real polynomials is twice the maximal degree of the summands.*

Proof. The reader is right to be surprised that these questions can be asked at all, that they are not utterly trivial. Of course, they *are* indeed easy, but as we shall have some more questions about polynomials, we shall take great care with answering them; in particular, we shall define the relevant terms carefully.

As usual, we denote by $\mathbb{R}[\mathbf{X}] = \mathbb{R}[X_1,\ldots,X_n]$ the set ('ring') of polynomials in n variables $X_1,\ldots,X_n$, and write $\mathbf{X}$ for the n-tuple $(X_1,\ldots,X_n)$. Let $\mathbb{N}^n$ be the set of n-tuples $\alpha = (\alpha_1,\ldots,\alpha_n)$, with each α_i a non-negative integer; we call these α *multidimensional exponents*. For $\alpha \in \mathbb{N}^n$ set $|\alpha| = \alpha_1 + \cdots + \alpha_n$ and $\mathbf{X}^\alpha = X_1^{\alpha_1} \ldots X_n^{\alpha_n}$; we call $|\alpha|$ the *degree* of the monomial $\mathbf{X}^\alpha$. Note that $\subset \mathbb{N}^n$ is endowed with an additive structure, and $\mathbf{X}^\alpha \mathbf{X}^\beta = \mathbf{X}^{\alpha+\beta}$.

The polynomials in $\mathbb{R}[\mathbf{X}]$ are precisely the real linear combinations of these monomials $\mathbf{X}^\alpha$: every polynomial $f \in \mathbb{R}[\mathbf{X}]$ has a canonical expansion

$$f = \sum_{\alpha \in A_f} c_\alpha \mathbf{X}^\alpha,$$

where A_f is a finite subset of $\mathbb{N}^n$ and for every $\alpha \in A_f$ the coefficient c_α is a non-zero real number. (In particular, f is the 0 polynomial iff A_f is the empty set.) The *degree* of a polynomial f is the maximal degree of a constituent monomial in the canonical expansion of f: $\deg f = \max\{|\alpha| : \alpha \in A_f\}$. (Thus the degree of the 0-polynomial is taken to be $-\infty$.) Furthermore, let D_f be the set of multidimensional exponents $\alpha \in A_f$ with $|\alpha| = \deg f$, i.e. let D_f be the set of exponents showing that $\deg f$ is (at least) as large as it is claimed to be.

(i) At long last, let us turn to the proof of the assertion that

$$\deg(fg) = \deg f + \deg g.$$

In proving this, we may assume that f and g are non-zero. Furthermore, as $\deg(fg) \leq \deg f + \deg g$ trivially holds, our task is to prove the converse inequality. Let

$$f = \sum_{\alpha \in A_f} c_\alpha \mathbf{X}^\alpha \quad \text{and} \quad g = \sum_{\beta \in A_g} d_\beta \mathbf{X}^\beta$$

be the canonical expansions of the non-zero polynomials $f, g \in \mathbb{R}[\mathbf{X}]$.

All we have to show is that there are $\alpha^* \in D_f$ and $\beta^* \in D_g$ such that they are the only elements of A_f and A_g that sum to $\alpha^* + \beta^*$: that will show that $\alpha^* + \beta^* \in A_{fg}$, in fact, the coefficient of $\mathbf{X}^{\alpha^*+\beta^*}$ in the expansion of fg is $c_{\alpha^*} d_{\beta^*}$. To this end, we simply choose α^* and β^* to be the maximal elements of D_f and D_g in an order compatible with addition.

For example, let $\alpha^* = (\alpha_1^*, \ldots, \alpha_n^*)$ be the maximal element of D_f in the so-called *lexicographic* order. Thus α^* is the element of D_f such that, first, α_1^* is maximal, then α_2^* is maximal, and so on, up to α_{n-1}^*. (We do not have any choice for α_n^* since that is determined by $|\alpha^*| = \deg f$ and the earlier α_i^*.) Spelling this out, if $\alpha \in D_f$ and $\alpha \neq \alpha^*$ then there is a k, $0 \leq k < n$, such that $\alpha_i = \alpha_i^*$ for $i \leq k$, and $\alpha_{k+1} < \alpha_{k+1}^*$. Define $\beta^* \in D_g$ analoguously.

Then the canonical expansion of fg contains the monomial

$$\mathbf{X}^{\gamma^*} = \mathbf{X}^{\alpha^*+\beta^*} = X_1^{\alpha_1^*+\beta_1^*} \cdots X_n^{\alpha_n^*+\beta_n^*}$$

with coefficient $c_{\alpha^*} d_{\beta^*}$, since for $\gamma^* = \alpha^* + \beta^* = (\alpha_1^* + \beta_1^*, \ldots, \alpha_n^* + \beta_n^*)$ the exponents $\alpha^* \in A_f$ and $\beta^* \in A_g$ are the unique elements of A_f and A_g with $\gamma^* = \alpha^* + \beta^*$. Hence $\deg(fg) \geq \deg f + \deg g$, as required.

(ii) Suppose $f = f_1^2 + \cdots + f_n^2$. Assume, as we may, that $f_1, \ldots, f_k$ have maximal degree d, say, where $1 \leq k \leq n$, so that $\deg f_i < d$ for $i > k$. Let α^* be the maximal element of $\bigcup_{i=1}^k D_{f_i}$ in the lexicographic order. If $\alpha^* \in D_{f_i}$ then, arguing as in part (i), the coefficient of $\mathbf{X}^{2\alpha^*}$ in f_i^2 is the square of the coefficent of $\mathbf{X}^{\alpha^*}$ in f_i, and if $\alpha^* \notin D_{f_i}$ then the coefficient of $\mathbf{X}^{2\alpha^*}$ in f_i^2 is 0. Hence the coefficient of $\mathbf{X}^{2\alpha^*}$ in f is strictly positive. $\qquad\square$

Notes. These assertions are just about the first (essentially trivial) facts one proves when introducing polynomial rings. Our Trinity undergraduates tended to assume that they were trivial, but then had trouble proving them.

72. Sums of Squares

(i) *A polynomial $f \in \mathbb{R}[X]$ is non-negative on $\mathbb{R}$ if and only if it is the sum of two squares: $f = g^2 + h^2$ for some $g, h \in \mathbb{R}[X]$.*
(ii) *If $f \in \mathbb{R}[X,Y]$ is non-negative on $\mathbb{R}^2$ then it does not follow that f is a sum of squares.*

Proof. (i) One of the implications is trivial: if f is the sum of squares then it is positive. To prove the converse, note that every non-zero polynomial $f \in \mathbb{R}[X]$ is of the form

$$f(X) = c \prod_1^k (X - r_i)^{\alpha_i} \prod_1^\ell ((X - s_j)^2 + t_j^2)$$

with $c \neq 0$, $r_1 < \cdots < r_k$ and $\alpha_i \geq 1$. Clearly, $f(x) \geq 0$ for every $x \in \mathbb{R}$ if and only if $c > 0$ and each exponent α_i is even. Hence, all we need is that

$$\prod_1^k ((X - s_j)^2 + t_j^2)$$

is a sum of two squares. This is an immediate consequence of the standard identity

$$(a^2 + b^2)(c^2 + d^2) = (ac + bd)^2 + (ad - bc)^2.$$

(ii) We claim that for $0 < c \leq 3$ the polynomial

$$f(X,Y) = 1 - cX^2Y^2 + X^4Y^2 + X^2Y^4$$

will do.
Indeed, by the AM–GM inequality,

$$\frac{1 + x^4y^2 + x^2y^4}{3} \geq x^2y^2$$

for all real x, y, so $f \geq 0$ on $\mathbb{R}^2$.

Suppose that, contrary to our assertion, $f = \sum_1^k f_i^2$ for some polynomials $f_1, \ldots, f_k \in \mathbb{R}[X, Y]$. Since f has degree 6, by Problem 71 each f_i has degree at most 3, i.e. each f_i is a linear combination of the monomials 1, X, Y, X^2, XY, Y^2, X^3, X^2Y, XY^2 and Y^3. In fact, a little thought tells us that most of these monomials cannot appear in any f_i. Indeed, if X^3 has coefficient a_i in f_i then the coefficient of X^6 in $\sum f_i^2$ is $\sum a_i^2$. Hence $a_i = 0$ for every i. Similarly, Y^3 does not appear in the f_i either.

Now, if b_i is the coefficient of X^2 in f_i then the coefficient of X^4 in $\sum f_i^2$ is $\sum b_i^2$; hence $b_i = 0$ for every i. Similarly, Y^2 does not appear in the f_i either. Next, we see that X and Y do not appear either. Consequently,

$$f_i = c_i + d_i XY + e_i' X^2Y + e_i'' XY^2.$$

But then the coefficient of X^2Y^2 in $f(X, Y)$ is $\sum_1^k d_i^2$, contradicting the fact that this coefficient is the negative number $-c$. $\qquad\square$

Notes. This simple exercise leads towards Hilbert's 17th problem: if $f \in \mathbb{R}[X_1, \ldots, X_n]$ is non-negative on $\mathbb{R}^n$, does it follow that f is a sum of squares of *rational* functions?

What we have just seen is that this is indeed true (even in the stronger form of being a sum of squares of *polynomials*) for $n = 1$ and is false in this stronger form for $n \geq 2$. In fact, the non-trivial part (ii) was already known to Hilbert in 1888, although his example was non-constructive. The first concrete example to appear in print seems to have been given by Motzkin in 1967: this is essentially the example we have given. Note that for $0 < c < 3$ this polynomial $1 - cX^2Y^2 + X^4Y^2 + X^2Y^4$ is strictly positive on $\mathbb{R}^2$. Numerous other examples have been contructed, e.g. by Choi and Lam, and Berg, Christensen and Jensen.

References

Berg, C., J.P.R. Christensen and C.U. Jensen, A remark on the multidimensional moment problem, *Math. Ann.* **243** (1979) 163–169.

Choi, M.D., Positive semidefinite biquadratic forms, *Linear Algebra Appl.* **12** (1975) 95–100.

Motzkin, T.S., The arithmetic–geometric inequality. In *Inequalities* (Proc. Sympos. Wright-Patterson Air Force Base, Ohio, 1965), pp. 205–224, Academic Press (1967).

73. Diagrams of Partitions

The number of partitions of n into p parts, with largest part q, is equal to the number of partitions of n into q parts, with largest part p. In particular, the number of partitions on n with maximal part p is the number of partitions of n into p parts.

Proof. This is a simple application of Ferrers diagrams (defined in the *Hint* to this problem), although we draw it the way Young tableaux are drawn, as an array of n top and left aligned cells (small squares). Let $\lambda = (\lambda_1, \ldots, \lambda_k)$ be a partition of n with Ferrers diagram D. Reflect D in its main diagonal (going from north-west to south-east): that gives a Ferrers diagram D^*, which defines a partition λ^* of n, the *conjugate* of λ. By definition, λ is the conjugate of λ^*.

Clearly, λ has length p and maximal part q if and only if λ^* has length q and its maximal part is p. Hence the two sets of partitions are indeed equinumerous. $\square$

Notes. This is an exceptionally simple application of Ferrers diagrams – perhaps *the* simplest. In the 19th century J.J. Sylvester called the correspondence $\lambda \to \lambda^*$ *the general theorem of reciprocity*. As Sylvester wrote in his long 1882 paper: '*The above proof of the theorem of reciprocity is due to Dr. Ferrers, the present head of Gonville and Caius' College, Cambridge. It possesses the double merit of having set the first example of graphical construction and of putting into salient relief the principle of correspondence, applied to the theory of partitions. It was never made public by its author, but first promulgated by myself in the* Lond. and Edin. Phil. Mag. *for 1853*'.

Norman Macleod Ferrers (1829–1903) read for the Mathematical Tripos at Gonville and Caius College, Cambridge, and was Senior Wrangler in 1851. Although he got a Fellowship at his college, he went to London to study law, and was called to the bar in 1855. Soon after that, he abandoned law and returned to Cambridge to study for the priesthood, and was ordained a priest in 1860. Nevertheless, he lectured on mathematics, and had the reputation of being the best lecturer in Cambridge. From 1880 until his death, he was Master of Gonville and Caius College.

Alfred Young (1873–1940) was also a Cambridge mathematician: he introduced Young diagrams and Young tableaux (Young diagrams filled with numbers) in his research into group theory. As an undergraduate at Clare College, he was 'only' Tenth Wrangler in 1895, but was considered to be the most original man of his year. Although he got a lectureship at Cambridge, later, like Ferrers before him, he studied for the priesthood and was ordained a priest in 1908. He remained a parish priest for the rest of his life, but after a substantial break he resumed lecturing on mathematics at Cambridge.

Reference

Sylvester, J.J., with insertions by Dr A. Franklin, Constructive theory of partitions, arranged in three acts, an interact and an exodion, *Amer. J. Math.* **5** (1882) 251–330.

74. Euler's Pentagonal Number Theorem

(i) *Denote by* $p_e(n)$ *the number of partitions of* n *into an even number of unequal summands, and by* $p_o(n)$ *the number of partitions of* n *into an odd number of unequal summands. Then* $p_e(n) - p_o(n) = 0$ *unless* $n = k(3k \pm 1)/2$, *in which case this difference is* $(-1)^k$:

$$p_e(n) - p_o(n) = \begin{cases} 1 & \text{if } n = k(3k \pm 1)/2 \text{ and } k \text{ is even,} \\ -1 & \text{if } n = k(3k \pm 1)/2 \text{ and } k \text{ is odd,} \\ 0 & \text{otherwise.} \end{cases}$$

(ii) *The infinite product* $\prod_{k=1}^{\infty}(1 - x^k)$ *has the following expansion:*

$$\prod_{k=1}^{\infty}(1 - x^k) = 1 + \sum_{k=1}^{\infty}(-1)^k (x^{k(3k-1)/2} + x^{k(3k+1)/2}). \tag{1}$$

Proof. (i) As usual, we identify a partition with its Ferrers diagram, using the 'Ferrers diagram style' rather than the 'Young tableau style'. This identification is unlikely to cause any confusion. We shall use D and λ for the same partition: as much of the time we care about the *shape* of the diagram, we tend to use D. Suppressing n, we write $\mathcal{D}$ for the set of (Ferrers diagrams of) the partitions of n with unequal summands, i.e. the set of diagrams in which each row is shorter than the one above it. Briefly, $\mathcal{D}$ consists of the set of *unequal partitions* of n. Denote by $\mathcal{D}_e \subset \mathcal{D}$ the subset of *even* partitions in $\mathcal{D}$, i.e. those with even length, and let $\mathcal{D}_o = \mathcal{D} \setminus \mathcal{D}_e$ be the set of *odd* partitions, i.e. those with odd length. Set $p_e(n) = |\mathcal{D}_e|$ and $p_o(n) = |\mathcal{D}_o|$. Thus, for $n = 6$ we have $\mathcal{D} = \{6, 51, 42, 321\}$, $\mathcal{D}_e = \{51, 42\}$ and $\mathcal{D}_o = \{6, 321\}$, showing that $p_e(6) = p_o(6) = 2$.

We shall define an appropriate involution F on $\mathcal{D}$, i.e. a map $F: \mathcal{D} \to \mathcal{D}$ such that F^2 is the identity. Our involution F will be such that if $F(D) \neq D$ then

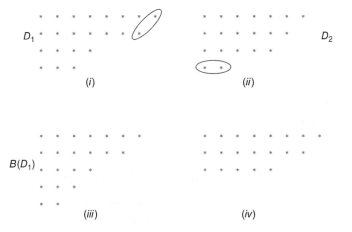

Figure 40 In the partition D_1 the side is $k = 2$ (and the line is $l = 3$), and in D_2 the line is $l = 2$ (and the side is $k = 3$). In $B(D_1)$ the base is the old side, and in $S(D_2)$ the side is the old base.

D and $F(D)$ belong to different sets of the partition $\mathcal{D}_e \cup \mathcal{D}_o$. This implies that the set of partitions that are not fixed points of F are divided equally between $\mathcal{D}_e$ and $\mathcal{D}_o$, so $p_e(n) - p_o(n) = |\mathcal{D}_e| - |\mathcal{D}_o|$ is the difference between the number of fixed points of F in $\mathcal{D}_e$ and the number of fixed points in $\mathcal{D}_o$. Furthermore, we shall be so lucky that there will be at most one fixed point; furthermore, for 'most' values of n there will be no fixed point.

To define F, we put together two different operators, B ('base operator') and S ('side operator'). Given a partition $D \in \mathcal{D}$, we denote by k the longest south-westerly line at the end of the diagram, starting in the top row, as in Figure 40(i), and call it the *side* of the diagram, and denote its last row by ℓ, as in Figure 40(ii), and call it the *base* of D. Also, we shall use the same letters for their cardinalities.

(a) If either $k < \ell - 1$ or $k = \ell - 1$ and k and ℓ do not meet, then we place the side below the base, as in Figure 40(iii) to obtain $B(D)$.

(b) If either $\ell < k$ or $\ell = k$ and k and ℓ do not meet, then we place the base next to the side, as in Figure 40(iv) to obtain $S(D)$.

Note that the cases (a) and (b) are exclusive: if one holds, the other does not. Also, B and D are inverses of each other in the sense that if $B(D)$ is defined then $S(B(D))$ is also defined and $S(B(D)) = D$; similarly, $B(S(D)) = D$. This

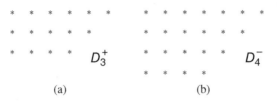

Figure 41 The diagrams D_3^+ and D_4^-. They are partitions of $3(3 \cdot 3 + 1)/2 = 15$ and $4(3 \cdot 4 - 1)/2 = 22$.

implies that for $D \in \mathcal{D}$ we may define our involution F as follows:

$$F(D) = \begin{cases} B(D) & \text{if } B \text{ is defined on } D, \\ S(D) & \text{if } S \text{ is defined on } D, \\ D & \text{otherwise.} \end{cases}$$

What are the fixed points of F? In other words, on which diagrams D is neither B nor S defined? If $k < \ell - 1$ then B is defined, if $\ell < k$ then S is defined; otherwise, if ℓ and k do not meet, either one or the other is defined, but not both. Hence, D is a fixed point if either $\ell = k + 1$ and the base ℓ and the side k meet, or else $\ell = k$ and the base ℓ and the side k meet. In each case, there is a unique diagram satisfying these conditions: we denote them by D_k^+ and D_k^-, as in Figures 41(a) and (b). Clearly, D_k^+ is the diagram of a partition of $n = k(3k+1)/2$ and D_k^- is a diagram of a partition of $n = k(3k-1)/2$. Hence, if $n \neq k(3k \pm 1)/2$ then F has no fixed point, if $n = k(3k-1)/2$ or $n = k(3k+1)/2$ then F has exactly one fixed point, either D_k^- or D_k^+. Furthermore, each of these 'fixed point' diagrams D_k^- and D_k^+ has k rows, so their partitions belong to $\mathcal{D}_e$ if k is even and to $\mathcal{D}_o$ if k is odd. This completes our proof of part (i).

(ii) In the expansion of $\prod_{k=1}^{\infty}(1 + x^k)$ the coefficient of x^n is the number of partitions of n into unequal parts. And what about the coefficient of x^n in the expansion of $\prod_{k=1}^{\infty}(1 - x^k)$? The same partitions are counted, but with signs: a partition of n into k unequal parts is counted with sign $(-1)^k$.   □

Notes. The wonderful proof of this famous result of Euler we have just presented is due to Fabian Franklin. He published it in Sylvester's 1882 paper and also in a paper of his in *Comptes Rendus*. (In Sylvester's paper it was claimed to be the first time Franklin's proof appeared in print.) This proof is often considered to be the first world-class contribution to mathematical research by an American.

In fact, Fabian Franklin (1853–1939) was born in the town of Eger, Hungary, famous for the glorious defence of its castle in 1552 by a handful of heroic defenders, including women, against the overwhelming force of the Ottoman Empire. (Ever since then Eger, its captain, István Dobó, and especially the

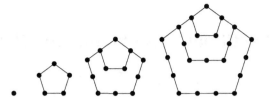

Figure 42 Representations of the pentagonal numbers 1, 5, 12 and 22.

'*women of Eger*', are the symbols of national heroism in defence of freedom.) It should be added that Franklin was actually thoroughly American, the favourite student of J.J. Sylvester during his seven years at Johns Hopkins. After his Ph.D., Franklin became Sylvester's assistant, and remained in close contact with Sylvester even when, at the age of 68, the latter went to Oxford to occupy the Savilian Chair of Geometry.

Returning to Euler's Pentagonal Theorem, Leonhard Euler (1707–1783) did not state his famous theorem as a result about even and odd partitions, but as a result about the expansion of an infinite product:

$$\prod_{k=1}^{\infty}(1-x^k) = \sum_{-\infty}^{\infty}(-1)^k x^{k(3k-1)/2} = 1 + \sum_{k=1}^{\infty}(-1)^k(x^{k(3k-1)/2} + x^{k(3k+1)/2}).$$

First he discovered this result empirically, but could not prove it: he found a proof some years later.

Finally, the name of the theorem is due to the fact that the numbers 1, 5, 12, ... (in general, $k(3k-1)/2$, for $k = 1, 2, \ldots$) are called *pentagonal numbers*, for the reason illustrated by Figure 42.

References

Euler, L., Découverte d'une loi tout extraordinaire des nombres par rapport à la somme de leurs diviseurs, *Bibliotheque Impartiale* **3** (1751) 10–31. Reprinted in *Opera Omnia Series I*, Vol. 2, pp. 241–253.

Euler, L., Demonstratio theorematis circa ordinem in summis divisorum observatum (1754–1755), *Novi Comment. Acad. Sci. Petropol.* **5** (1760) 75–83. Reprinted in *Opera Omnia Series I*, Vol. 2, pp. 390–398.

Euler, L., Evolutio producti infiniti $(1-x)(1-xx)(1-x^3)(1-x^4)(1-x^5)(1-x^6)$ etc. in seriem simplicem, *Acta Acad. Sci. Imp. Petropol.* **I** (1780) 47–55. Reprinted in *Opera Omnia Series I*, vol. 3, pp. 472–479.

Franklin, F., Sur le développement du produit infini $(1-x)(1-x^2)(1-x^3)\cdots$, *Comptes Rendus* **82** (1881) 448–450.

Sylvester, J.J., with insertions by Dr A. Franklin, Constructive theory of partitions, arranged in three acts, an interact and an exodion, *Amer. J. Math.* **5** (1882) 251–330.

75. Partitions – Maximum and Parity

Let $p_{o,e}(n)$ be the number of partitions of n into unequal parts in which the largest part is odd and the number of parts is even, and define $p_{e,o}(n)$, $p_{e,e}(n)$ and $p_{o,o}(n)$ analogously. Then

$$p_{o,e}(n) - p_{e,o}(n) = \begin{cases} 1 & \text{if } n = k(3k-1)/2 \text{ and } k \text{ is even,} \\ -1 & \text{if } n = k(3k+1)/2 \text{ and } k \text{ is odd,} \\ 0 & \text{otherwise,} \end{cases}$$

and

$$p_{o,o}(n) - p_{e,e}(n) = \begin{cases} 1 & \text{if } n = k(3k-1)/2 \text{ and } k \text{ is odd,} \\ -1 & \text{if } n = k(3k+1)/2 \text{ and } k \text{ is even,} \\ 0 & \text{otherwise.} \end{cases}$$

Proof. As in Franklin's proof of Euler's Pentagonal Theorem in the previous problem, let F be the involution on the set $\mathcal{D}_n$ of Ferrers diagrams of partitions of n into unequal parts. Furthermore, let $\mathcal{D}_{o,e} \subset \mathcal{D}_n$ be the subset of diagrams with largest part odd and even length, and define $\mathcal{D}_{e,o}$, $\mathcal{D}_{e,o}$ and $\mathcal{D}_{e,o}$ analogously, so that $p_{o,e}(n) = |\mathcal{D}_{o,e}|$, etc. As was shown in Problem 74, F has at most one fixed point: this fixed point is D_k^- if $n = k(3k-1)/2$, and it is D_k^+ if $n = k(3k+1)/2$. If $n \neq k(3k \pm 1)/2$ then F has no fixed point on $\mathcal{D}_n$. Also, D_k^- has maximal part $2k-1$ and length k, and D_k^+ has maximal part $2k$ and length k.

By definition, apart from its fixed point, our involution gives a one-to-one correspondence between $\mathcal{D}_{o,e}$ and $\mathcal{D}_{e,o}$, and between $\mathcal{D}_{o,o}$ and $\mathcal{D}_{e,e}$. Since D_k^- is in $\mathcal{D}_{o,e}$ if k is even and in $\mathcal{D}_{o,o}$ if k is odd, and D_k^+ is in $\mathcal{D}_{e,o}$ if k is odd and in $\mathcal{D}_{e,e}$ if k is even, the required formulas follow. $\square$

Notes. In this problem we have studied the standard division of $\mathcal{D}_n$ into four parts, with cardinalities $p_{o,e}(n)$, $p_{e,o}(n)$, $p_{o,o}(n)$ and $p_{e,e}(n)$. As we have seen,

$p_{o,e}(n)$ and $p_{e,o}(n)$ are just about equal, and so are $p_{o,o}(n)$ and $p_{e,e}(n)$. It remains to find out how large the differences like $|p_{o,e}(n) - p_{o,o}(n)|$ can be. Is it true that there is a constant C such that $|p_{o,e}(n) - p_{o,o}(n)| \le C$ if n is large enough? If this is not true, about how large can this difference be as a function of n?

76. Periodic Cellular Automata

Let G be a finite graph in which every degree is odd. Then every majority bootstrap percolation on G is periodic, with period 1 or 2.

Proof. Let $f = (f_t)_0^\infty$ be a majority bootstrap percolation on G. Since G is finite, f is periodic with some period $k \geq 1$. Suppose that, contrary to the assertion, $k \geq 3$. Assume, as we may, that $f_{t+k} = f_t$ for every $t \geq 0$. To reduce clutter, write v_t for the state $f_t(v)$ of a vertex v at time t.

Using the Kronecker δ function (defined by $\delta(x, y) = 1$ if $x = y$ and $\delta(x, y) = 0$ otherwise), the majority update rule tells us that

$$\sum_{u \in \Gamma(v)} \delta(u_{t-1}, v_t) \geq \sum_{u \in \Gamma(v)} \delta(u_{t-1}, \omega)$$

for $\omega = 0, 1$, with equality (trivially) for $\omega = v_t$ and strict inequality otherwise (i.e. if $\omega = 1 - v_t$). In particular,

$$\sum_{u \in \Gamma(v)} \left(\delta(u_{t-1}, v_t) - \delta(u_{t-1}, v_{t-2}) \right) \geq 0 \tag{1}$$

for all $v \in V(G)$ and $t \geq 0$, with equality if and only if $v_t = v_{t-2}$. Needless to say, the suffices are taken modulo k: in particular, $v_k = v_0$, $v_{k-1} = v_{-1}$, etc.

Now, as $k \geq 3$, there is a vertex $v \in V$ such that $v_{t-2} \neq v_t$ for some time t; in particular, inequality (1) is strict for this pair (v, t). Consequently,

$$L = \sum_{t=1}^{k} \sum_{v \in V} \sum_{u \in \Gamma(v)} \left(\delta(u_{t-1}, v_t) - \delta(u_{t-1}, v_{t-2}) \right) > 0. \tag{2}$$

Note that in this triple sum every adjacent pair (u, v) of vertices occurs twice: once in the order first u and then v and once when we take v first and then u. Let us take these two occurrences together so that we are summing over the

220

edges uv:

$$L = \sum_{t=1}^{k} \sum_{uv \in E(G)} \left(\delta(u_{t-1}, v_t) - \delta(u_{t-1}, v_{t-2}) + \delta(v_{t-1}, u_t) - \delta(v_{t-1}, u_{t-2}) \right).$$

To complete the proof we shall interchange the order of summation, and then sum the terms separately. First, by interchanging the order of summation, we find that

$$L = \sum_{uv \in E(G)} \sum_{t=1}^{k} \left(\delta(u_{t-1}, v_t) - \delta(u_{t-1}, v_{t-2}) + \delta(v_{t-1}, u_t) - \delta(v_{t-1}, u_{t-2}) \right).$$

Second, since $\delta(u, v) = \delta(v, u)$ for all $u, v \in V$, we have the following identities:

$$\sum_{t=1}^{k} \delta(u_{t-1}, v_t) = \sum_{t=1}^{k} \delta(u_{t-2}, v_{t-1}) = \sum_{t=1}^{k} \delta(v_{t-1}, u_{t-2})$$

and

$$\sum_{t=1}^{k} \delta(u_{t-1}, v_{t-2}) = \sum_{t=1}^{k} \delta(u_t, v_{t-1}) = \sum_{t=1}^{k} \delta(v_{t-1}, u_t).$$

Consequently, $L = 0$, contradicting (2) and so completing our proof. □

Notes. The result in this problem was proved by Goles and Olivos in 1981; a little later, it was rediscovered by Poljak and Sûra. In fact, Poljak and Sûra proved much more than we have stated. Also, after trivial modifications, the proof above gives much more: we may take cellular automata with weighted edges in which each vertex can be in finitely many states. The update is again given by the majority, with tie-breaks allowed.

More formally, here is the variant proved by Poljak and Sûra.

Let V be a finite set and w a symmetric real function on V; thus, $w :$ $V \times V \to \mathbb{R}$, with $w(x, y) = w(y, x)$. Let $f_0, f_1, \ldots$ be functions mapping V into $\{0, 1, \ldots, p\}$; we call each f_t a configuration, and $f_t(v)$ the state of v at time t. A sequence $(f_t)_0^\infty$ of configurations is a generalized majority bootstrap configuration if the update rule is the following:

$$f_{t+1}(v) = \max\left\{ i : \sum_{u:f_t(u)=i} w(u, v) \geq \sum_{u:f_t(u)=j} w(u, v) \text{ for every } j \right\}.$$

In words, at time $t + 1$ a vertex v goes into the highest value state which is at least as frequent among the neighbours as any other state. Here the frequency is measured using the weights given by w, and the highest value is used only to ensure that ties are broken somehow. Note that if w takes only the values 0 and 1 and $w(x, x) = 0$ for every x then $\{uv : u, v \in V, \; w(u, v) = 1\}$ is the edge set of a graph; if, furthermore, every degree is odd and $p = 1$ then we get back the original setup. The *period* of a majority bootstrap percolation $f = (f_t)_0^\infty$ is again the smallest natural number s such that $f_{t+s} = f_t$ if t is large enough.

Poljak and Sûra proved that *the period of such a generalized majority bootstrap percolation is* 1 *or* 2 *as well.* A moment's thought tells us that the proof above carries over to this general case as well, *mutatis mutandis.*

References

Goles, E. and J. Olivos, Comportement périodique des fonctions á seuil binaires et applications, *Discrete Appl. Math.* **3** (1981) 93–105.

Poljak, S. and M. Sûra, On periodical behaviour in societies with symmetric influences, *Combinatorica* **3** (1983) 119–121.

77. Meeting Set Systems

Let $\mathcal{A} = \{A_1,\ldots,A_n\} \subset S^{(\leq k)}$ be such that no element of S is contained in more than d of the A_i. Then the probability that a p-random subset X_p of S meets $\mathcal{A}$ is at most

$$\left(1 - q^k\right)^{n/d},$$

where $q = 1 - p$.

Proof. We prove the assertion by induction on n. Since for $n = 1$ the assertion is trivial, with equality, let us turn to the induction step. Assuming, as we may, that $A_1 = [k] = \{1,\ldots,k\}$, let E_ℓ be the event that ℓ is the smallest element of $X_p \cap A_1$. Thus the events $E_1,\ldots,E_k$ are disjoint, and $\bigcup_1^k E_\ell$ is the event that $X_p \cap A_1 \neq \emptyset$. Also, let

$$\mathcal{A}_\ell = \{A_i \setminus [\ell - 1] : \ell \notin A_i\}.$$

Note that X_p meets $\mathcal{A}$ if and only if for some ℓ, $1 \leq \ell \leq k$, the event E_ℓ holds and X_p meets $\mathcal{A}_\ell$. Also, if E_ℓ holds then $|\mathcal{A}_\ell| \geq n - d$, since at most d sets A_i contain ℓ. Hence, by the induction hypothesis,

$$\mathbb{P}\left(X_P \text{ meets } \mathcal{A}_\ell \mid E_\ell\right) \leq \left(1 - q^k\right)^{(n-d)/d}.$$

Finally, since

$$\mathbb{P}\left(\bigcup_1^k E_\ell\right) = \sum_{\ell=1}^k \mathbb{P}(E_\ell) = 1 - q^k,$$

it follows that

$$\mathbb{P}(X_p \text{ meets } \mathcal{A}) \leq \left(1 - q^k\right)\left(1 - q^k\right)^{(n-d)/d} = \left(1 - q^k\right)^{n/d},$$

as claimed. $\square$

Note. This is a lemma Newman used in one of his papers.

Reference

Newman, D.J., Complements of finite sets of integers, *Michigan Math. J.* **14** (1967) 481–486.

78. Dense Sets of Reals – An Application of the Baire Category Theorem

Let P be a set of points in the plane $\mathbb{R}^2$ such that the set

$$\{x/y : (x,y) \in P, y \neq 0\}$$

is dense in $\mathbb{R}$. Then there is an $\alpha \in \mathbb{R}$ such that

$$\{x + \alpha y : (x,y) \in P\}$$

is also dense in $\mathbb{R}$.

Proof. For a natural number n and a rational q, define

$$G_{n,q} = \{\alpha : |x + \alpha y - q| < 1/n \quad \text{for some } (x,y) \in P\}.$$

The set $G_{n,q}$ is open, since it is the union of some open sets; furthermore, it is dense in $\mathbb{R}$, since $(q - x)/y \in G_{n,q}$ for all $(x,y) \in P$. Hence, by the Baire Category Theorem, the intersection G of all these sets $G_{n,q}$ is also dense in $\mathbb{R}$; *a fortiori*, $\alpha \in G$ for some real α. For such an α, the set $\{x + \alpha y : (x,y) \in P\}$ is trivially dense in $\mathbb{R}$. $\qquad\qquad\square$

Notes. The result above is due to McMullen, who proved it to solve the following problem of Pach.

Let P be a set of points in the plane that meets every unit disc. Show that there is a line ℓ such that the orthogonal projection of P into ℓ is dense in ℓ.

In fact, McMullen deduced from his theorem the following stronger result inspired by this problem of Pach.

Let P be a set of points in the plane such that $d(z,P) = o(||z||)$ as $||z|| \to \infty$. Then there is a line ℓ such that the orthogonal projection of P into ℓ is dense in ℓ.

Curiously, when the problem was first published in *The Monthly*, it was

225

attributed to Beck, Galvin and Pach, but in the issue containing the solution (and extension) given by McMullen, Pach was given as the sole author of the problem.

References

Beck, J., F. Galvin and J. Pach, Problems and Solutions: Advanced Problems: 6421, *Amer. Math. Monthly* **90** (1983) 134.

Pach, J. and C. McMullen, Problems and Solutions: Solutions of Advanced Problems: 6421, *Amer. Math. Monthly* **91** (1984) 589.

79. Partitions of Boxes

An n-dimensional combinatorial box cannot be partitioned into fewer than 2^n non-trivial sub-boxes.

Proof. Following the *Hint*, let $A = A_1 \times \cdots \times A_n$ be an n-dimensional combinatorial box, and let $\mathcal{B}$ be its partition into non-trivial sub-boxes. Thus $\mathcal{B}$ is a collection of disjoint non-trivial boxes and $A = \bigcup_{B \in \mathcal{B}} B$. We call a sub-box $C = C_1 \times \cdots \times C_n$ *odd* if it has an odd number of elements, i.e. if $|C_i|$ is odd for every i. For a sub-box B, let O_B be the set of odd sub-boxes of A meeting B in an odd sub-box; thus O_A is the set O of all odd sub-boxes.

The claim is a consequence of the following two observations.

(i) If B is a non-trivial sub-box then

$$|O_B| = |O|/2^n.$$

Indeed, for any non-empty finite set, half of its subsets are odd, i.e. have an odd number of elements. Hence, if $\emptyset \neq B_i \neq A_i$, then half of the subsets of A_i are odd, and a quarter are odd *and* meet B_i in an odd set.

(ii) If $C \in O$ then $|C| = \sum_{B \in \mathcal{B}} |C \cap B|$ is odd, so at least one of the summands is odd; consequently,

$$\bigcup_{B \in \mathcal{B}} O_B = O.$$

Putting these two observations together, we find that

$$|O| = \left| \bigcup_{B \in \mathcal{B}} O_B \right| \leq |\mathcal{B}| \, |O|/2^n,$$

so $|\mathcal{B}| \geq 2^n$, as required. $\qquad\square$

Note. The beautiful result above was conjectured by Kearnes and Kiss, and proved by Alon, Bohman, Holzman and Kleitman.

References

Alon, N., T. Bohman, R. Holzman and D.J. Kleitman, On partitions of discrete boxes, *Discrete Math.* **257** (2002) 255–258.

Kearnes, K.A. and E.W. Kiss, Finite algebras of finite complexity, *Discrete Math.* **207** (1999) 89–135.

80. Distinct Representatives

Let $A_1, \ldots, A_n$ be finite sets satisfying

$$\sum_{1 \leq i < j \leq n} \frac{|A_i \cap A_j|}{|A_i||A_j|} < 1.$$

Then the sets A_i have distinct representatives: there are distinct elements $a_1, \ldots, a_n$ such that $a_i \in A_i$ for every i.

Proof. There are $\prod_1^n |A_i|$ maps $f : [n] \to \bigcup_1^n A_i$ with $f(i) \in A_i$: our aim is to show that one of these maps is injective. The number of maps with $f(i) = f(j)$ is

$$|A_i \cap A_j| \prod_{k \neq i, j} |A_k| = \left(\prod_1^n |A_i| \right) \frac{|A_i \cap A_j|}{|A_i||A_j|},$$

and so, recalling the condition, the number of non-injective maps is at most

$$\left(\prod_1^n |A_i| \right) \sum_{1 \leq i < j \leq n} \frac{|A_i \cap A_j|}{|A_i||A_j|} < \prod_1^n |A_i|,$$

completing the proof. $\square$

Note. The observation above and its natural proof are due to Václav Chvátal.

Reference

V. Chvátal, Problem 2309, *Amer. Math. Monthly* **79** (1972) 775.

81. Decomposing a Complete Graph: The Graham–Pollak Theorem – Take One

A complete graph on n vertices cannot be decomposed into n − 2 complete bipartite graphs.

Proof. Let us proceed as suggested in the *Hint.* Suppose for a contradiction that the complete graph on $[n] = \{1, 2, \ldots, n\}$ is decomposed into $r \leq n - 2$ complete bipartite graphs $G_1, G_2, \ldots, G_r$, with G_i having classes U_i and W_i. Set $P_i = \sum_{j \in U_i} X_j$ and $Q_i = \sum_{j \in W_i} X_j$, so that

$$X_1 X_2 + X_1 X_3 + X_1 X_4 + \cdots + X_{n-1} X_n = P_1 Q_1 + P_2 Q_2 + \cdots + P_r Q_r,$$

i.e.

$$\sum_{i<j} X_i X_j = \sum_{i=1}^{r} P_i Q_i.$$

Since $r \leq n - 2$, the set of $r + 1 \leq n - 1$ linear equations

$$P_1 = P_2 = \cdots = P_r = X_1 + X_2 + \cdots + X_n = 0$$

has a non-trivial solution $(c_1, \ldots, c_n)$ over the reals:

$$P_i(c_1, \ldots, c_n) = 0 \qquad \text{for } i = 1, \ldots, r,$$

and

$$c_1 + c_2 + \cdots + c_n = 0.$$

But this gives us the contradiction

$$0 < c_1^2 + \cdots + c_n^2 = (c_1 + \cdots + c_n)^2 - 2 \sum_{i<j} c_i c_j$$

$$= -2 \sum_{i=1}^{r} P_i(c_1, \ldots, c_n) Q_i(c_1, \ldots, c_n) = -2 \sum_{i=1}^{r} 0 \cdot Q_i(c_1, \ldots, c_n) = 0,$$

completing our proof. □

230

Notes. This beautiful and fundamental theorem was proved by Graham and Pollak in 1971. The simple and elegant proof given above was found by Tverberg in 1982. I consider it the most 'natural' of the many proofs of this fundamental result, since the existence of a graph decomposition is equivalent to the polynomial identity

$$X_1 X_2 + X_1 X_3 + X_1 X_4 + \cdots + X_{n-1} X_n = P_1 Q_1 + P_2 Q_2 + \cdots + P_r Q_r.$$

Later we shall see two further simple proofs.

References

Graham, R.L., and H.O. Pollak, On the addressing problem for loop switching, *Bell System Tech. J.* **50** (1971) 2495–2519.

Tverberg, H., On the decomposition of K_n into complete bipartite graphs, *J. Graph Theory* **6** (1982) 493–494.

82. Matrices and Decompositions: The Graham–Pollak Theorem – Take Two

A complete graph on n vertices cannot be decomposed into $n - 2$ complete bipartite graphs.

Proof. (i) Let H be the union of a complete bipartite graph with vertex classes U and W, and a set I of isolated vertices, so that H has vertex set $V = U \cup W \cup I$. For notational convenience, we take $V = [n] = \{1, \ldots, n\}$. Let B be the adjacency matrix of H, and let $C = (c_{ij})_{i,j=1}^{n}$ be the matrix with

$$c_{ij} = \begin{cases} 1 & \text{if } i \in U, j \in W, \\ 0 & \text{otherwise.} \end{cases}$$

Then C has rank 1, since its ith row $(c_{ij})_{j=1}^{n}$ is either identically 0 (if $i \notin U$) or 1 if $j \in W$ and 0 otherwise. Also, $B = C + C^t$, where C^t is the transpose of C, so $B = 2C + (C^t - C) = 2C - D$. Thus B is the sum of $2C$, a matrix of rank 1, and $D = C^t - C$, which is antisymmetric.

(ii) Let $(G_i)_1^r$ be a decomposition of the complete graph K_n with vertex set $[n]$ into r complete bipartite graphs, $G_1, \ldots, G_r$. The adjacency matrix of K_n is $A = J - I$, where J is the identically 1 matrix and I is the identity. Writing B_i for the adjacency matrix of G_i, we have

$$A = \sum_{i=1}^{r} B_i,$$

since the G_i decompose K_n. Hence, by (i),

$$A = \sum_{i=1}^{r} \left(C_i + D_i \right) = \sum_{i=1}^{r} C_i + D,$$

where the C_i have rank 1 and the D_i, and so D, are antisymmetric. As $A = J - I$,

232

we see that

$$J - \sum_{i=1}^{r} C_i = I + D.$$

Finally, the left-hand side has rank at most $r + 1$, since it is the sum of $r + 1$ matrices of rank 1, and the right-hand side has rank n, since D, being antisymmetric, has only imaginary eigenvalues. Consequently, $r \geq n - 1$, proving our assertion.   □

Notes. This proof of the Graham–Pollak theorem was given by D.J. Kleitman (writing as G.W. Peck) in 1984: it can be viewed as the matrix formulation of the 1982 proof by Tverberg using quadratic forms we gave earlier.

References

Graham, R.L. and H.O. Pollak, On the addressing problem for loop switching, *Bell System Tech. J.* **50** (1971) 2495–2519.

Peck, G.W., A new proof of a theorem of Graham and Pollak, *Discrete Math.* **49** (1984) 327–328.

83. Patterns and Decompositions: The Graham–Pollak Theorem – Take Three

A complete graph on n vertices cannot be decomposed into $n - 2$ complete bipartite graphs.

Proof. Proceeding as in the original formulation of this problem, for a given N and a decomposition of K_n, there are at most $(nN)^{r+1}$ patterns, while the number of maps $f : [n] \to [N]$ is N^n. Hence, if $r \leq n - 2$ and $N > n^{r+1}$ then there are two maps, f and g, say, with the same pattern.

For $1 \leq i \leq n$, define $c_i = f(i) - g(i)$. Then the sequence $(c_i)_1^n$ is not identically 0, but

$$\sum_{i \in U_k} c_i = 0 \quad \text{for } 1 \leq k \leq r, \quad \text{and} \quad \sum_1^n c_i = 0.$$

Consequently,

$$0 < \sum_{i=1}^n c_i^2 = \left(\sum_{i=1}^n c_i\right)^2 - 2\sum_{i<j} c_i c_j$$

$$= -2\sum_{k=1}^r \left(\sum_{i \in U_k} c_i\right)\left(\sum_{j \in W_k} c_j\right) = 2\sum_{k=1}^r 0 \cdot \left(\sum_{j \in W_k} c_j\right) = 0.$$

This contradiction tells us that r cannot be at most $n - 2$. ☐

Notes. This lovely proof of the Graham–Pollak theorem, given by Vishwanathan in 2013, is again very close to the 1982 proof by Tverberg using quadratic forms we have already reproduced. In this proof a non-trivial solution of the $r + 1$ equations $\sum_{i \in U_k} c_i = 0$, $1 \leq k \leq r$, and $\sum_1^n c_i = 0$ is deduced by counting, so the proof loses its algebraic flavour.

234

References

Graham, R.L. and H.O. Pollak, On the addressing problem for loop switching, *Bell System Tech. J.* **50** (1971) 2495–2519.

Vishwanathan, S., A counting proof of the Graham–Pollak theorem, *Discrete Math.* **313** (2013) 765–766.

84. Six Concurrent Lines

Let P_1, P_2, P_3 and P_4 be four points on a circle. For $1 \le i < j \le 4$ let ℓ_{ij} be the line through the midpoint of the segment $P_i P_j$ and perpendicular to the line $P_h P_k$, where $\{i, j, h, k\} = \{1, 2, 3, 4\}$. Then the six lines ℓ_{ij} are concurrent.

Proof. We may assume that our points are unit complex numbers, $z_1, \ldots, z_4$, say. Perhaps a little surprisingly, the point our six lines go through is *twice* the arithmetic mean of the z_i. Thus, all we have to check is that the line through $(z_1 + z_2)/2$ and $(z_1 + \cdots + z_4)/2$, which has direction $z_3 + z_4$, is perpendicular to $z_3 - z_4$, i.e. that the ratio $(z_3 - z_4)/(z_3 + z_4)$ is pure imaginary. This is trivial to check:

$$\frac{z_3 - z_4}{z_3 + z_4} = \frac{(z_3 - z_4)(\bar{z}_3 + \bar{z}_4)}{|z_3 + z_4|^2} = \frac{z_3 \bar{z}_4 - \bar{z}_3 z_4}{|z_3 + z_4|^2},$$

and the difference of a complex number and its conjugate is pure imaginary. □

Notes. This exercise is really for students who have just encountered complex numbers: I first heard it from István Reiman when I was about twelve, and I enjoyed doing it. The assertion is of course just the tip of the iceberg, as any good text on elementary geometry will tell us.

Reference

Yaglom, I.M., *Complex Numbers in Geometry*. Translated from the Russian by Eric J.F. Primrose, Academic Press (1968).

85. Short Words – First Cases

For $k \geq 2$ and $m \geq 1$, write $n_k(m)$ for the minimal cardinality of an unavoidable set over an alphabet of k letters in which each word has length at least m. Show that $n_k(1) = k$, $n_k(2) = \binom{k+1}{2}$ and $n_2(3) = 4$.

Proof. Let $A = \{1, 2, \ldots, k\}$ be our alphabet, and let X be an unavoidable set of words of length at least m with $|X| = n_k(m)$. A word in X which has length greater than m can be replaced by a factor of length m, so we may assume that every word in X has length m.

(i) The case $m = 1$. Since for $1 \leq i \leq k$ the words i^ℓ with ℓ large have a factor in X, $i \in X$. Hence, $|X| \geq k$. Also, $X = \{1, 2, \ldots, k\}$ is trivially unavoidable.

(ii) The case $m = 2$. For $1 \leq i \leq j \leq k$, long words of the type $(ij)^2 = ijij \cdots ij$ have a factor in X, so at least one of ij and ji is in X. Consequently, $n_k(2) = |X| \geq \binom{k+1}{2}$.

Conversely, the set

$$X_0 = \{ij : 1 \leq i \leq j \leq k\}$$

is clearly unavoidable since if a word $w = w_1 w_2 \ldots w_\ell$ avoids X_0 then $w_1 > w_2 > \cdots > w_\ell$, so $\ell \leq k$. Since $|X_0| = \binom{k+1}{2}$, we find that $n_k(2) = \binom{k+1}{2}$, as claimed.

(iii) Finally, we take $k = 2$ and $m = 3$, and determine $n_2(3)$. First, the set $X_0 = \{111, 222, 112, 212\}$ is unavoidable. Indeed, as 111 and 112 are in X, this set is unavoidable if $\{11, 222, 212\}$ is. Now, if $i1j$ is a factor of a word w then w contains either 11 or 212 Hence X_0 is indeed unavoidable, so $n_2(3) \leq 4$.

On the other hand, for both $122122 \ldots 122$ and $112112 \ldots 112$ to have factors in our unavoidable set X, we need two 'mixed' words in X. Also,

237

X must contain 111 and 222 as well since otherwise $11 \ldots 1$ or $22 \ldots 2$ avoids X. Thus $|X| \geq 4$, and so $n_2(3) = 4$, as required.   □

Notes. Unavoidable sets were studied by Higgins and Saker: in the next problem we return to the topic with a more general result.

References

Higgins, P.M., The length of short words in unavoidable sets, *Int. J. Algebra Comput.* **21** (2011) 951–960.

Higgins, P.M. and C.J. Saker, Unavoidable sets, *Theoret. Comput. Sci.* **359** (2006) 231–238.

86. Short Words – The General Case

Let $k \geq 2$ and $m \geq 1$, and write $n_k(m)$ for the minimal cardinality of an unavoidable set over an alphabet A of k letters in which each word has length at least m. Then

$$k^m/m \leq n_k(m) \leq k^m. \tag{1}$$

Proof. The set A^m of k^m words of length m is unavoidable, so the second inequality in (1) holds. To prove the first, let X be an unavoidable set of $n_k(m)$ words, with each word of length at least m. Then for some $t \geq m$, no word of length at least t avoids X. Also, if we replace a word $x \in X$ of length greater than m by one of its factors of length m, the new set is also unavoidable. Hence, we may assume that X consists of words of length m, i.e. $X \subset A^m$.

If $w \in A^m$ then $w^t = ww \cdots w$, the periodic word of length m, does not avoid X. Since X consists of words of length (at most) m, w^2 does not avoid X either. Briefly, no word in the set of squares

$$S = \{w^2 : w \in A^m\}$$

avoids X. Since every word $x \in X$ is the factor of at most m words in S,

$$|S| \leq m|X|,$$

i.e.

$$n_k(m) = |X| \geq |S|/m = k^m/m,$$

completing our proof. □

Notes. Let us state the result above in its dual form. Write $m_k(n)$ for the maximal integer such that every unavoidable set of cardinality n consists of words of length at least m. Then

$$\log_k n \le m_k(n) \le \log_k n + \log_k \log_k n + 1. \qquad (2)$$

The first inequality here is precisely the second inequality in (1), but the second in (2) is a little weaker than the first in (1). Indeed, for $k = 2$ and $n = 3, 4$ this follows from $n_2(2) = 3$ and $n_2(3) = 4$ in the previous problem, so we may assume that either $k = 2$ and $n \ge 5$ or else $k \ge 3$.

We have $m \le k^{m-1}$ for $m \ge 1$ and $k \ge 2$, so $m = m_k(n) > \log_k n + \log_k \log_k n + 1$ and (1) lead to the contradiction

$$n \ge k^m/m > n(k \log_k n)/(\log_k n + \log_k \log_k n + 1) \ge n,$$

since

$$(k - 1) \log_k n \ge \log_k \log_k n + 1$$

if $k = 2$ and $n \ge 5$, or $k \ge 3$.

The results in this problem (in the dual form) were proved by Higgins and Saker; later a simpler proof was given by Higgins. The proof above is even simpler.

References

Higgins, P.M., The length of short words in unavoidable sets, *Int. J. Algebra Comput.* **21** (2011) 951–960.
Higgins, P.M., and C.J. Saker, Unavoidable sets, *Theoret. Comput. Sci.* **359** (2006) 231–238.

87. The Number of Divisors

Write $d(n)$ for the number of divisors of a natural number n. Then $d(n) \leq 2\sqrt{n}$ and $d(n) = n^{o(1)}$.

Proof. (i) If $r \geq 1$ is a divisor of n then $n = rs$ for some $s \geq 1$, and at least one of r and s is at most $\sqrt{n}$. Hence, $d(n) \leq 2\sqrt{n}$.

(ii) Writing $p_1 < p_2 < \cdots$ for the sequence of primes, so that $p_1 = 2$, $p_2 = 3$, and so on, let $n = p_1^{a_1} p_2^{a_2} \cdots p_k^{a_k}$ be the prime factorization of n. Then $d(n) = (a_1 + 1) \cdots (a_k + 1)$, since the set of divisors of n is

$$\{p_1^{b_1} p_2^{b_2} \cdots p_k^{b_k} : \quad 0 \leq b_i \leq a_i;\ i = 1, \ldots, k\}.$$

To construct the prime factorization of n we start with 1 and include the prime factors of n one by one. At each step, we follow by how much the logarithm of the current number and the logarithm of its number of divisors increase.

Suppose a prime p appears in the current factorization with exponent $a - 1$, where $a \geq 1$. Let us add another factor p to the prime factorization, so that the exponent of p is increased from $a - 1$ to a. Then, $\log d(n)$ is increased by $\log((a+1)/a)$ and $\log n$ by $\log p$. In this sequence each pair (p, a) can only occur at most once, and for every $\varepsilon > 0$ there are only finitely many pairs (p, a) with $\log((a+1)/a)$ greater than $\varepsilon \log p$. Thus after including all of these first to obtain n_ε and continuing to some $n > n_\varepsilon$, we have $\log d(n) \leq \log d(n_\varepsilon) + \varepsilon \log(n/n_\varepsilon)$. Hence, very crudely, $d(n) \leq d(n_\varepsilon)n^\varepsilon$.

This tells us that $d(n) =)(n^\varepsilon)$ for every $\varepsilon > 0$, so $d(n) = n^{o(1)}$, as claimed. $\square$

Notes. This is a rather weak form of a basic result about divisors. The particularly elegant proof above I have learnt from Paul Balister.

In 1907, Wigert made use of the Prime Number Theorem to prove the exact result that

$$\limsup \frac{(\log_2 d(n))(\log \log n)}{\log n} = 1,$$

i.e. the maximal order of $\log_2 d(n)$ is $\log n / \log \log n$. As reported in Chapter 18 of Hardy and Wright, Ramanujan proved this without assuming the Prime Number Theorem.

References

Hardy, G.H. and E.M. Wright, *An Introduction to the Theory of Numbers*, Sixth edition. Revised by D.R. Heath-Brown and J.H. Silverman. With a foreword by Andrew Wiles. Oxford University Press (2008).

Wigert, S., Sur l'ordre de grandeur du nombre de diviseurs d'un entier, *Ark. Mat.* **3** (1906/1907) 1–9.

88. Common Neighbours

Every graph of order n and size m has two vertices with at least $\ell = \lfloor 4m^2/n^3 \rfloor$ common neighbours.

Proof. Suppose that the assertion is false and there is a graph G with n vertices and m edges, in which any two vertices have at most $\ell - 1$ common neighbours. Let $(d_i)_1^n$ be the degree sequence of G, so that $d = \frac{1}{n} \sum_{i=1}^n d_i = 2m/n$ is the average degree of the vertices. Counting paths of length two (i.e. triples of vertices a, b, c with b joined to both a and c) in two different ways, we find that

$$\sum_{i=1}^n \binom{d_i}{2} \le (\ell - 1)\binom{n}{2}.$$

Hence, since the binomial function $\binom{x}{2}$ is convex,

$$n\binom{d}{2} \le (\ell - 1)\binom{n}{2},$$

i.e.

$$\ell - 1 \ge \frac{2m}{n}\left(\frac{2m}{n} - 1\right)/(n-1).$$

However, this implies the contradiction that

$$\ell > 4m^2/n^3. \qquad \square$$

Note. This is a frequent step in an argument in extremal combinatorics.

89. Squares in Sums

Let A be a set of n integers such that $A + A$ contains the first m squares. Then $n \geq m^{2/3+o(1)}$.

Proof. Define a graph G on A as follows. For each k, $1 \leq k \leq m$, pick two numbers, $a, b \in A$, with $a + b = k^2$, and join a to b by an edge. Then G has n vertices and at least m edges. By the result in the previous problem, G has two vertices, a and $b < a$, say, with common neighbours $c_1, \ldots, c_\ell$, where $\ell = \lfloor 4m^2/n^3 \rfloor$. With $a+c_i = x_i^2$ and $b+c_i = y_i^2$, we have $a-b = (x_i+y_i)(x_i-y_i)$, with $i = 1, \ldots, \ell$. Consequently $a - b$ can be written in (at least) ℓ different ways as a product of two positive numbers: $a-b = r_i s_i$, where $1 \leq r_i < \sqrt{a - b} < m$, for $i = 1, \ldots, \ell$. Thus $a - b < m^2$ has at least 2ℓ divisors. By the result in Problem 87,

$$m^2/n^3 < 2\ell \leq (m^2)^{o(1)} = m^{o(1)},$$

so $n \geq m^{2/3+o(1)}$, as claimed. $\qquad\square$

90. Extension of Bessel's Inequality – Bombieri and Selberg

Let $\varphi_1, \varphi_2, \ldots, \varphi_n$ and f be vectors in a real Hilbert space H with inner product $(\ ,\ )$, and set $s_i = \sum_{j=1}^{n} |(\varphi_i, \varphi_j)|$. Then

$$\|f\|^2 \geq \sum_{i=1}^{n} (f, \varphi_i)^2 / s_i. \tag{1}$$

Proof. Given constants $\xi_1, \xi_2, \ldots, \xi_n$, we have

$$0 \leq \left\| f - \sum_{i=1}^{n} \xi_i \varphi_i \right\|^2 = \left(f - \sum_{i=1}^{n} \xi_i \varphi_i, f - \sum_{j=1}^{n} \xi_j \varphi_j \right)$$

$$= \|f\|^2 - 2 \sum_{i=1}^{n} \xi_i (f, \varphi_i) + \sum_{i,j=1}^{n} \xi_i \xi_j (\varphi_i, \varphi_j).$$

Since

$$\xi_i \xi_j \leq \frac{1}{2} (\xi_i^2 + \xi_j^2),$$

this implies that

$$\|f\|^2 \geq 2 \sum_{i=1}^{n} \xi_i (f, \varphi_i) - \sum_{i,j=1}^{n} \xi_i^2 (\varphi_i, \varphi_j) = 2 \sum_{i=1}^{n} \xi_i (f, \varphi_i) - \sum_{i=1}^{n} \xi_i^2 s_i.$$

Setting $\xi_i = (f, \varphi_i)/s_i$, we get (1). □

Notes. The simple inequality above is from a short paper of Bombieri on the large sieve: Bombieri attributes the inequality and its proof to Selberg. With the aid of this inequality, Bombieri gives a simpler proof of the large sieve inequality he proved with Davenport, resulting in an even better constant.

References

Bombieri, E., A note on the large sieve, *Acta Arith.* **18** (1971) 401–404.
Bombieri, E. and H. Davenport, Some inequalities involving trigonometrical polynomi-
    als, *Ann. Scuola Norm. Sup. Pisa Cl. Sci.* (3) **23** (1969) 223–241.

91. Equitable Colourings

Let S be a finite set of points in the plane $\mathbb{R}^2$. Then there is a red–blue colouring of the points of S such that every (coordinate) line is equitably coloured.

Proof. We prove the assertion by induction on n. For $n = 1$ (and, in fact, for 2, 3 and 4) the assertion is trivial, so assume that $n > 1$ and it holds for sets with at most $|S| - 1$ points. Let X be the set of lines parallel to the x-axis and going through at least one point of S, and let Y be the analogous set of lines for the y-axis. Let G_S be the bipartite graph with vertex partition (X, Y), in which a line $\ell \in X$ is joined to a line $m \in Y$ if they meet in a point of S. We shall distinguish two cases.

(i) Suppose that G_S contains a cycle $\ell_1 m_1 \ell_2 m_2 \cdots \ell_k m_k$, with the lines ℓ_i and m_i meeting in $B_i \in S$ and m_i and ℓ_{i+1} (with ℓ_{k+1} taken to be ℓ_1) in $R_i \in S$, as illustrated in Figure 43. Colour the points B_i blue and the points R_i red, and delete these points from S to obtain S'. By the induction hypothesis, S' has a suitable red–blue colouring: the two colourings together give a suitable colouring of S.

(ii) Suppose G_S contains no cycles, as in Figure 44. Then pick a line $\ell \in X$

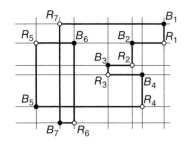

Figure 43 A cycle in G_S and its derived colouring.

247

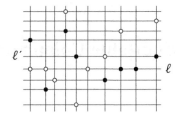

Figure 44 A colouring of G_S without a cycle, starting from ℓ and then ℓ'.

and colour it equitably. Next, take all the lines (in Y) meeting ℓ (in a point of S): each of these lines has one point which has already been coloured, so this colouring can be extended to an equitable colouring of the entire line. Continue in this way: at each stage, take all lines that we haven't yet coloured entirely but contain a coloured point. As G_S has no cycle, each of these lines contains *exactly one* coloured point, so we can extend this (very partial) colouring to an equitable colouring of the entire line. The process stops when we have no new line meeting any of the lines we have already coloured. Deleting the set of points on these lines, we are done by the induction hypothesis.   □

Notes. This is one of the simplest problems about equitable colourings. It is natural to ask whether one can impose the additional condition that the total number of points coloured red differs from the total number of points coloured blue by at most one. Not to spoil the fun, we do not give the answer.

92. Scattered Discs

Let $D_1, \ldots, D_n$ be unit discs (in the plane) with centres $c_1, \ldots, c_n$, $n \geq 3$, such that no line meets more than two of them. Then

$$\sum_{1 \leq i < j \leq n} \frac{1}{d_{ij}} < \frac{n\pi}{4},$$

where $d_{ij} = d(c_i, c_j)$ is the distance between c_i and c_j.

Proof. Consider one of the centres, say c_i. For a disc D_j, $1 \leq j \leq n$, $j \neq i$, let C_{ij} be the double cone bounded by the two tangents of D_j from c_i, as in Figure 45. Writing φ_{ij} for the half-angle of C_{ij} at its centre c_i, so that $\varphi_{ij} < \pi/2$, we have

$$\frac{1}{d_{ij}} = \sin \varphi_{ij} < \varphi_{ij}.$$

As no line meets more than two of the discs $D_1, \ldots, D_n$, the $n-1$ double cones $C_{i1}, C_{i2}, \ldots, C_{in}$, with centre c_i, are disjoint, except for their common

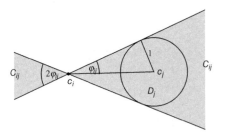

Figure 45 The double cone C_{ij} and its angle φ_{ij}.

249

centre. Hence

$$\sum_{\substack{1 \le j \le n \\ j \ne i}} \varphi_{ij} < \frac{\pi}{2}$$

for every i, so

$$\sum_{1 \le i < j \le n} \frac{1}{d_{ij}} < \sum_{1 \le i < j \le n} \varphi_{ij} = \frac{1}{2} \sum_{i=1}^{n} \sum_{\substack{j=1 \\ j \ne i}}^{n} \varphi_{ij} < \frac{n\pi}{4},$$

as claimed.   □

Notes. This simple problem bears a superficial (very superficial!) resemblence to Pólya's 'Orchard Problem'. Suppose we stand in a forest with circular tree trunks of radius ε in which no two trees are centred closer than unit distance apart. How far can we see? If the centres are arranged in such a way that for every $\varepsilon > 0$ there is a constant d that no matter where we stand, we cannot see further than distance d then our forest is said to be *dense*. Chris Bishop, Yuval Peres, Noga Alon, and others have proved results about dense forests.

References

Alon, N., Uniformly discrete forests with poor visibility, *Combin. Probab. Comput.* **27** (2018) 442–448.

Bishop, C.J., A set containing rectifiable arcs QC-locally but not QC-globally, *Pure Appl. Math. Q.* **7** (2011) 121–138.

93. East Model

Let $V(n)$ be the set of configurations in the East Model that can be created from the empty configuration by East processes if at each step we have at most n occupied sites. For $n \geq 1$, define

$$A(n) = \max\{x : \{x\} \in V(n)\}$$

and

$$B(n) = \max\{x : x \in X \ \text{for some} \ X \in V(n)\}.$$

Then $A(n) = 2^{n-1}$ and $B(n) = 2^n - 1$.

Proof. As suggested in our *Hint*, we shall make use of the following two facts.

Fact (i) Every process is reversible: if in a process in $V(n)$ we can change $X \in V(n)$ into $Y \in V(n)$ through a sequence of steps in $V(n)$, then in $V(n)$ we can change Y into X through a sequence of steps in $V(n)$. For example, writing 124 for the set $\{1, 2, 4\}$,

$$\emptyset \to 1 \to 12 \to 2 \to 23 \to 234 \to 24 \to 124 \to 14 \to 4$$

is a process in $V(3)$, and so is its reverse

$$4 \to 14 \to 124 \to 24 \to 234 \to 23 \to 2 \to 12 \to 1 \to \emptyset.$$

Fact (ii) If $X \in V(n)$ and $Y \in V(n-|X|)$ then $X \cup (x+Y) \in V(n)$ for every $x \in X$, where, as usual, $x + Y = \{x + y : y \in Y\}$. For example, $\{12, 17\} \in V(5)$ and $\{3, 5, 6\} \in V(3)$, so

$$\{12, 17\} \cup \{15, 17, 18\} = \{12, 15, 17, 18\} \in V(5),$$

and

$$\{12, 17\} \cup \{20, 22, 23\} = \{12, 17, 20, 22, 23\}.$$

Here Fact (i) is immediate, since the state of site i can be changed *either way* if $i - 1$ is occupied. Also, Fact (ii) holds since if $\emptyset \to Y_1 \to Y_2 \to \cdots \to Y_\ell = Y$ is an East process in $V(n - |X|)$ then for $X_i = X \cup (x + Y_i)$, the sequence $X \to X_1 \to \cdots \to X_\ell = X \cup (x + Y)$ is a process in $V(n)$.

After these remarks, we shall determine $A(n)$ and $B(n)$ simultaneously, by induction on n. We have seen that $A(1) = B(1) = 1$, $A(2) = 2$ and $B(2) = 3$, so we turn to the induction step. Assume that $n \geq 2$ and $A(k) = 2^{k-1}$ and $B(k) = 2^k - 1$ for $1 \leq k \leq n$; our task is to prove that $A(n + 1) = 2^n$ and $B(n) = 2^{n+1} - 1$.

Lower bounds. (a) Set $X = Y = \{2^{n-1}\}$ so that $X \in V(n) \subset V(n + 1)$ and $Y \in V(n) = V(n + 1 - |X|)$. Hence, by Fact (ii),

$$\{2^{n-1}, 2^n\} = X \cup \{2^{n-1} + Y\} \in V(n + 1).$$

By Fact (i), in $V(n)$ there is an East process $\{2^{n-1}\} \to Z_1 \to \cdots \to Z_\ell = \emptyset$, so the sequence $\{2^{n-1}, 2^n\} \to Z_1 \cup \{2^n\} \to \cdots \to Z_\ell \cup \{2^n\} = \{2^n\}$ is a process in $V(n + 1)$. Hence $\{2^n\} \in V(n + 1)$, i.e. $A(n + 1) \geq 2^n$.

(b) As we have just seen, $X = \{2^n\} \in V(n+1)$; also, by the induction hypothesis, there is a set $Y \in V(n)$ such that $\max Y = 2^n - 1$. Then, by Fact (ii), $2^n + Y \in V(n + 1)$, so $B(n + 1) \geq \max(2^n + Y) = 2^n + (2^n - 1) = 2^{n+1} - 1$.

Upper bounds. (a) Suppose that, contrary to what we wish to prove, $\{x_0\} \in V(n + 1)$ for some $x_0 \geq 2^n + 1$. By Fact (i), there is an East process $\{x_0\} \to X_1 \to X_2 \to \cdots \to X_\ell = \emptyset$ in $V(n + 1)$. We may assume that $X_i \subset \{1, \ldots, x_0\}$ for every i. Let $m = \min\{i : x_0 \notin X_i\}$ and set $X_i' = X_i \cap \{1, \ldots, x_0 - 1\}$. With this notation, $\emptyset \to X_1' \to X_2' \to \cdots \to X_m'$ is also an East process in $V(n)$. Clearly, $x_0 - 1 \in X_m'$ since $x_0 \in X_{m-1}$ but $x_0 \notin X_m$. Hence $\max X_m' = x_0 - 1 \geq 2^n$, contradicting our induction hypothesis that $B(n) = 2^n - 1$. This contradiction proves that $A(n + 1) \leq 2^n$, and so $A(n + 1) = 2^n$.

(b) Finally, before we prove that $B(n+1) \leq 2^{n+1} - 1$, we define another function on the East Model,

$$C(m) = \max_{X \in V(m)} \min X.$$

We claim that $C(n + 1) \leq 2^n$. To see this, let $\emptyset \to X_1 \to X_2 \to \cdots \to X$ be a process in $V(n+1)$ such that $\min X = C(n+1)$. If in this process we change the state of every site greater than $C(n + 1)$ from occupied to unoccupied, we get a lazy process in $C(n+1)$, ending in $\{C(n+1)\}$. Hence $C(n+1) \leq A(n+1) = 2^n$, as claimed.

The upper bound on $B(n + 1)$ is an easy consequence of this. Indeed, every configuration in $V(n + 1)$ has at most n occupied sites greater than $C(n + 1)$

and so certainly has at most n occupied sites greater than 2^n. Hence, by the induction hypothesis,

$$B(n + 1) \leq C(n + 1) + B(n) \leq 2^n + (2^n - 1) = 2^{n+1} - 1,$$

completing our proof.   □

Notes. The last argument tells us more about the East Model. Indeed, by expanding the last display, we find that

$$2^{n+1} - 1 = B(n + 1) \leq C(n + 1) + C(n) + \cdots + C(2) + C(1)$$
$$\leq 2^n + 2^{n-1} + \cdots + 2 + 1 = 2^{n+1} - 1.$$

Consequently, $C(k) = 2^{k-1}$ for every $k \geq 1$.

Furthermore, $B(n)$ is attained on a *unique* configuration, namely on the configuration $X = \{x_1, \ldots, x_n\} \in V(n)$ such that $x_1 = 2^{n-1}$, $x_2 = 2^{n-1} + 2^{n-2}$, $\ldots$, $x_n = 2^{n-1} + 2^{n-2} + \cdots + 1 = 2^n - 1$.

The East Model is one of the simplest members of the large family of *kinetically constrained models* used to explain some features of dynamics of gases – see the tiny selection of papers below. Chung, Diaconis and Graham were the first to define and find the values of the functions $A(n)$ and $B(n)$: this evaluation is an introductory result in their paper in which they study the entropy of the East Model, i.e. the cardinality of $V(n)$. They prove upper and lower bounds of the form $2^{\binom{n}{2}} n! c^n$ for various constants c with $0 < c < 1$.

References

Aldous, D., and P. Diaconis, The asymmetric one-dimensional constrained Ising model: Rigorous results, *J. Statist. Phys.* **107** (2002) 945–975.

Chung, F., P. Diaconis and R. Graham, Combinatorics for the East model, *Adv. Appl. Math.* **27** (2001) 192–206.

Faggionato, A., F. Martinelli, C. Roberto and C. Toninelli, The East model: Recent results and new progresses, *Markov Process. Related Fields* **19** (2013) 407–452.

Martinelli, F., R. Morris and C. Toninelli, Universality results for kinetically constrained spin models in two dimensions, *Comm. Math. Phys.* **369** (2019) 761–809.

Valiant, P., Linear bounds on the North–East model and higher-dimensional analogs, *Adv. Appl. Math.* **33** (2004) 40–50.

94. Perfect Triangles

Show that there are five perfect triangles, i.e. triangles in which the length of every side is an integer and whose area is equal to the length of its perimeter. The triples of side lengths are as follows: $(5, 12, 13)$, $(6, 8, 10)$, $(6, 25, 29)$, $(7, 15, 20)$, $(9, 10, 17)$.

Proof. Let $a \le b \le c$ be the side lengths of a perfect triangle, and set $s = \frac{a+b+c}{2}$. By Heron's formula, the area of this triangle is $\sqrt{s(s-a)(s-b)(s-c)}$, so

$$2s = \sqrt{s(s-a)(s-b)(s-c)}.$$

To reduce the clutter in our formulae, set $\ell = s - c$, $m = s - b$ and $n = s - a$, so $\ell \le m \le n$, $s = \ell + m + n$, and

$$4(\ell + m + n) = \ell m n. \tag{1}$$

The time has come to find the solutions in ℓ, m, n and so a, b, c. First, $\ell \ge 4$ is impossible, since then we would have

$$4(\ell + m + n) \le 12n < 16n \le \ell m n,$$

contradicting (1). The remaining three cases, $\ell = 1, 2, 3$, are taken care of one by one.

First, if $\ell = 1$ then (1) becomes

$$(m - 4)(n - 4) = 20,$$

giving that (m, n) is $(5, 24)$, $(6, 14)$ or $(8, 9)$.

If $\ell = 2$ then (1) can be written as

$$(m - 2)(n - 2) = 8,$$

so (m, n) can be $(3, 10)$ or $(4, 6)$.

254

Finally, if $\ell = 3$ then (1) tells us that

$$(3m - 4)(3n - 4) = 52.$$

This equation has no solutions in positive integers, since the divisors of 52 are 1, 2, 4, 13, 26 and 52, so $3m - 4$ should be at most 4. However, this is not the case since $m \geq \ell = 3$.

This completes the proof since the triples we have found for (ℓ, m, n) give the claimed triples for (a, b, c). $\qquad\square$

Notes. This simple problem is very old indeed. B. Yates posed it as Question 2019 in *The Lady's and Gentleman's Diary* in 1865, pp. 49–50. Independently of this, W.A. Whitworth and D. Biddle published it in 1904 in the *Educational Times*, pp. 54–56 and 62–63.

In fact, earlier Carl Friedrich Gauss had studied a similar but more difficult problem. Suppose that the sides and the radius of the circumscribed circle of a triangle are integers. In a letter to H.C. Schumacher on 21st October, 1847, Gauss gave the following parametrization of the three sides of such a triangle:

$$4abfg(a^2 + b^2), \qquad |4ab(f + g)(a^2 f - b^2 g)|, \qquad 4ab(a^2 f^2 + b^2 y^2).$$

95. A Triangle Inequality

Let a, b and c be the sides, and Δ the area of a triangle. Then

$$a^2 + b^2 + c^2 \geq 4\sqrt{3}\Delta.$$

Proof. By Heron's formula, $\Delta = (s(s-a)(s-b)(s-c))^{1/2}$, where s is the half-perimeter of the triangle. Hence, squaring the inequality we wish to prove and substituting $s = (a+b+c)/2$, our task is to show that

$$(a^2 + b^2 + c^2)^2 \geq 3(a+b+c)(-a+b+c)(a-b+c)(a+b-c),$$

i.e.

$$(a^2 + b^2 + c^2)^2 \geq 6(a^2b^2 + b^2c^2 + c^2a^2) - 3(a^4 + b^4 + c^4).$$

Setting $A = a^2$, $B = b^2$ and $C = c^2$, expanding the left-hand side, rearranging the terms, and dividing by 2, our task becomes the trivially true inequality

$$(A - B)^2 + (B - C)^2 + (C - A)^2 \geq 0,$$

completing our proof. □

Notes. Although this inequality was published by Weitzenböck in 1919, and was sharpened by Finsler and Hadwiger in 1937, it became known when in the 1940s Pedoe rediscovered it and popularized it. This very simple inequality has a good many proofs: in particular, Pedoe deduced it from an exercise on orthogonal projections, and Finsler and Hadwiger derived it from a geometric calculation. Here we have given a pedestrian (mindless?) proof, taking heed of Einstein's saying that 'chalk is cheaper than grey matter'.

References

Finsler, P. and H. Hadwiger, Relationen im Dreieck *Comment. Math. Helv.* **10** (1937) 316–326.

Pedoe, D., Orthogonal projections of triangles, *Math. Gaz.* **25** (1941) 224.

Weitzenböck, R., Über eine Ungleichung in der Dreieckgsgeometrie, *Math. Zeit.* **5** (1919) 137–146.

96. An Inequality for Two Triangles

We are given two triangles: one with sides a, b, c and area Δ, and another with sides a', b', c' and area Δ'. Then

$$a'^2(-a^2 + b^2 + c^2) + b'^2(a^2 - b^2 + c^2) + c'^2(a^2 + b^2 - c^2) \geq 16\Delta\Delta'. \quad (1)$$

Proof. We shall use the identities $c^2 = a^2 + b^2 - 2ab(\cos \gamma)$ and $\Delta = ab(\sin \gamma)/2$, where γ is the angle opposite side c, and the analogues of these identities for the other triangle. Using these, the left-hand side of (1) can be written as

$$2(a^2b'^2 + b^2a'^2) - (a^2 + b^2 - c^2)(a'^2 + b'^2 - c'^2)$$
$$= 2(a^2b'^2 + b^2a'^2) - 4aa'bb' \cos \gamma \cos \gamma'$$
$$= 2(a^2b'^2 + b^2a'^2) - 4aa'bb'(\cos(\gamma - \gamma') - \sin \gamma \sin \gamma')$$
$$\geq (ab' - ba')^2 + 4aa'bb' \sin \gamma \sin \gamma'$$
$$\geq 16\Delta\Delta',$$

completing our proof. ☐

Notes. Applying the inequality with $a' = b' = c' = 1$, we get Weitzenböck's inequality for one triangle, as in Problem 95.

The inequality itself is due to Daniel Pedoe, but the elegant proof we have given is from E.H. Neville, whom Pedoe must have known at Cambridge, and whom he credits in the 1941 paper below. J.E. Littlewood considered the proof particularly beautiful, and remembered it many years later, when he told me about it. (As I wanted to put this vaguely remembered result into the present collection of problems, I searched for a reference *for ages*: it is a miracle that I was successful just in time.) That Littlewood remembered the result and its proof must have had much to do with the person of Neville, who came Second Wrangler in 1909, the very last time the Order of Merit was decided in the Mathematical Tripos. Since then there has been no strict order, so no Senior Wrangler.

In 1909 there was much *national* interest as to who would be the last Senior Wrangler: in the cut-throat competition P.J. Daniell came top, and E.H. Neville was followed by Louis Mordell, who went on to be a great number theorist. (Although there is no doubt that Mordell became the greatest of the three, all three became excellent mathematicians.) According to legend, Mordell was sent to Cambridge to make sure that the last Senior Wrangler was an American: as Mordell said later, he blotted his copy-book and came 'only' third.

References

Pedoe, D., An inequality connecting any two triangles, *Math. Gaz.* **25** (1941) 310–311.
Pedoe, D., An inequality for two triangles, *Math. Gaz.* **26** (1942) 397–398.
Weitzenböck, R., Über eine Ungleichung in der Dreiecksgeometrie, *Math. Zeit.* **5** (1919) 137–146.

97. Random Intersections

Let $\mathcal{A} \subset \mathcal{P}_n$ and set

$$\mathcal{J} = \mathcal{J}(\mathcal{A}) = \{A \cap B : A, B \in \mathcal{A}\}.$$

Then for $0 < p < 1$ we have

$$\mathbb{P}_{p^2}(\mathcal{J}) \geq \mathbb{P}_p(\mathcal{A})^2.$$

Proof. Set $r = \mathbb{P}_p(\mathcal{A})$, and let X_p and Y_p be independent p-random subsets of $[n]$, so that $r = \mathbb{P}(X_p \in \mathcal{A})$. Furthermore, $X_p \cap Y_p$ is a p^2-random subset of $[n]$ since the events $\{1 \in X_p \cap Y_p\}, \{2 \in X_p \cap Y_p\}, \ldots, \{n \in X_p \cap Y_p\}$ are independent, and each has probability p^2. Consequently, the p^2-probability of the family $\mathcal{J}$ of intersections is the probability that $X_p \cap Y_p$ belongs to $\mathcal{J}$. Hence

$$\mathbb{P}_{p^2}(\mathcal{J}) = \mathbb{P}(X_p \cap Y_p \in \mathcal{J}) \geq \mathbb{P}(X_p \in \mathcal{A} \text{ and } Y_p \in \mathcal{A})$$
$$= \mathbb{P}(X_p \in \mathcal{A}) \mathbb{P}(Y_p \in \mathcal{A}) = r^2,$$

as claimed. □

Notes. This inequality was used by Ellis and Narayanan in their beautiful paper in which they proved the old conjecture of Peter Frankl that if $\mathcal{A} \subset \mathcal{P}_n$ is a *symmetric* 3-wise intersecting family then $|\mathcal{A}| = o(2^n)$. (Thus $A \cap B \cap C \neq \emptyset$ for all $A, B, C \in \mathcal{A}$ and the automorphism group of $\mathcal{A}$ is transitive on $[n]$, i.e. for all $1 \leq i < j \leq n$ there is a permutation of $[n]$ that maps i into j, and maps every set in $\mathcal{A}$ into a set in $\mathcal{A}$.) More precisely, Ellis and Narayanan proved this for $p = 1/2$: sitting in a seminar Narayanan was giving on their proof of Frankl's conjecture, Paul Balister gave this extension and the lovely proof above.

The proof trivially carries over to the following extension. Let $\mathcal{A}_1, \ldots, \mathcal{A}_k$ be families of subsets of $[n]$, and $0 < p_1, \ldots, p_k < 1$. Then

$$\mathcal{J} = \{A_1 \cap \ldots A_k : A_i \in \mathcal{A}_i\}$$

satisfies

$$\mathbb{P}_{p_k} \geq \prod_{i=1}^{k} \mathbb{P}_{p_i}(\mathcal{A}_i).$$

Reference

Ellis, D., and B. Narayanan, On symmetric 3-wise intersecting families, *Proc. Amer. Math. Soc.* **145** (2017) 2843–2847.

98. Disjoint Squares

Let $\mathcal{F} = \{Q_1,\ldots,Q_n\}$ be a family of standard unit squares in $\mathbb{R}^2$ with their union $A = \bigcup_{i=1}^{n} Q_i$ having area $|A| > 4k$. Then $\mathcal{F}$ contains a subcollection of $k + 1$ pairwise-disjoint unit squares.

Show also that if the area of A is $4k$, we cannot guarantee $k + 1$ disjoint squares.

Proof. For $(x, y) \in 2\mathbb{Z}^2$, let $R_{(x,y)}$ be the standard 2×2 square in $\mathbb{R}^2$ with centre (x, y), i.e. with vertices $(x + 1, y + 1), (x - 1, y + 1), (x - 1, y - 1), (x + 1, y - 1)$. For each square $R_{(x,y)}$ with $A \cap R_{(x,y)} \neq \emptyset$, let $A_{(x,y)}$ be this intersection translated into $R_{(0,0)}$ through $-(x, y)$:

$$A_{(x,y)} = \left(A \cap R_{(x,y)}\right) - (x, y).$$

Since translation preserves area,

$$\sum_{(x,y)} |A_{(x,y)}| = \sum_{(x,y)} |A \cap R_{(x,y)}| = |A| > 4k.$$

As $R_{(0,0)}$ has area 4, it has an interior point (u, w) which is also the interior point of $k + 1$ sets

$$A_{(x_1,y_1)}, A_{(x_2,y_2)}, \ldots, A_{(x_{k+1},y_{k+1})}.$$

Here the points (x_i, y_i) are $k + 1$ different lattice points of $2\mathbb{Z}^2$. Since $A_{(x_i,y_i)} = A \cap R_{(x_i,y_i)} - (x_i, y_i)$ and A is the union of our squares $Q_1,\ldots,Q_n$, there are squares $Q_{n_1},\ldots,Q_{n_{k+1}}$ such that $(u, w) \in Q_{n_i} - (x_{n_i}, y_{n_i})$ for every i, i.e. $(u_i, w_i) \in Q_{n_i}$, where $u_i = u + x_{n_i}$ and $w_i = w + y_{n_i}$.

As two unit squares containing different points of $2\mathbb{Z}^2$ in their interiors are disjoint, the squares $Q_{n_1},\ldots,Q_{n_{k+1}}$ are disjoint, completing our proof.

To see the second part, simply take k disjoint standard 2×2 squares, divide

262

each into four standard unit squares, sharing only some sides and vertices. Then no two of these four squares are disjoint, so we are done. □

Notes. Trivially, the first assertion is true for open squares as well. However, the second part is a little different: total area $4k$ is not sufficient for standard open squares. However, for every $\varepsilon > 0$ there are k groups of four squares, almost as above, but *slightly* intersecting, such that the union has area at least $4k - \varepsilon$, and no $k + 1$ of the squares are pairwise disjoint.

This exercise is a special case of the following problem posed by Tibor Radó in 1928. If the area of a family of standard squares in $\mathbb{R}^2$ is 1, what area can we guarantee for a subset of disjoint squares? What we have proved above is that if the squares are *congruent* then we can guarantee area at least $1/4$. In fact, this result was first proved by Sokolin in 1940 and, independently, by Richard Rado (no relation!) in 1949, with a follow-up paper in 1951, in which several variants of the original problem were considered. In 1958, Norlander also rediscovered the result in this exercise.

The method above is applicable to families of *congruent* standard cubes in $\mathbb{R}^d$: it tells us that if their union has unit volume then there is a subcollection of pairwise disjoint cubes of volume at least $1/2^d$.

Tibor Radó conjectured that the condition that the squares are congruent is not needed: if the area of a family of standard squares in $\mathbb{R}^2$ is 1, then there is a subset of disjoint squares whose union has area at least $1/4$. Amazingly, 45 years later, Ajtai constructed a counterexample to this conjecture. Ajtai's example was tweaked by Bereg, Dumitrescu and Joang to show that we cannot guarantee more than area $1/4 - 1/384$. Nevertheless, the problem concerning the maximum of the area we can guarantee is still wide open.

References

Ajtai, M., The solution of a problem of T. Radó, *Bull. Acad. Polon. Sci. Sér. Sci. Math. Astronom. Phys.* **21** (1973) 61–63.

Bereg, S., A. Dumitrescu and M. Jiang, On covering problems of Rado, *Algorithmica* **57** (2010) 538–561.

Norlander, G., A covering problem (in Swedish, with English summary), *Nordisk Mat. Tidskr.* **6** (1958) 29–31.

Rado, R., Some covering theorems (I), (II), *Proc. London Math. Soc.* **51** (1949) 241–264 and **53** (1951) 243–267.

Radó, T., Sur un problème relatif à un théorème de Vitali, *Fundam. Math.* **11** (1928) 228–229.

Sokolin, A., Concerning a problem of Radó, *C. R. (Doklady) Acad. Sci. URSS (N.S.)* **26** (1940) 871–872.

99. Increasing Subsequences – Erdős and Szekeres

(i) *Let $a = (a_i)_1^n$ be a sequence of $n = pq + 1$ real numbers, where p and q are natural numbers. Then a has either a strictly increasing subsequence of length $p + 1$ or a decreasing subsequence of length $q + 1$.*

(ii) *For all natural numbers p, q and $n = pq \geq 1$, there is a sequence $a = (a_i)_1^n$ of natural numbers that contains neither an increasing subsequence of length $p + 1$ nor a decreasing subsequence of length $q + 1$.*

Proof. (i) Suppose a does not contain a strictly increasing subsequence of length $p + 1$. Colour a term a_i with the maximal length of a strictly increasing subsequence ending in a_i. [For example, a_i is coloured 1 if $a_j \geq a_i$ for every j, $1 \leq j < i$.] Then every a_i is coloured with one of $1, 2, \ldots, p$, so for some k at least $\lceil n/p \rceil = q + 1$ of the terms get the same colour k. Clearly, if $i < j$ and a_i and a_j get the same colour then $a_i \geq a_j$, so the terms of colour k form a decreasing subsequence of length at least $q + 1$.

(ii) For every i, $1 \leq i \leq n$, there are unique integers j, r with $0 \leq j \leq p - 1$ and $1 \leq r \leq q$ such that $i = jq + r$. Set $a_i = (j + 1)q + 1 - r$. The sequence we obtain, $q, q - 1, \ldots, 1; 2q, 2q - 1, \ldots, q + 1; \ldots; pq, pq - 1, \ldots, (p-1)q + 1$, will clearly do, see Figure 46 for $p = 3$ and $q = 5$. $\quad\square$

Notes. Part (i) is (a slightly stronger version of) a very simple result from one of the most popular papers on combinatorics ever, the fundamental paper of Erdős and Szekeres from 1935. The original version of that simple result was Exercise 2(iv) in *Coffee time in Memphis*. Actually, we may change a subsequence of equal terms to a strictly decreasing subsequence without changing any other order, so the original version implies this version instantly. But here we are juggling with two essentially trivial results. Later we shall return to the main result from the Erdős–Szekeres paper.

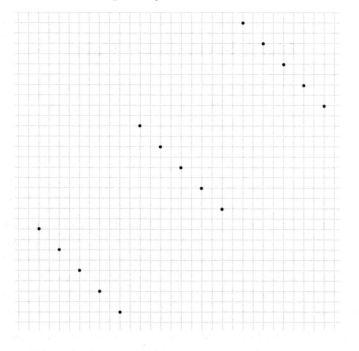

Figure 46 The sequence 5, 4, . . . , 1; 10, 9, . . . , 6; 15, 14, . . . , 11.

Reference

Erdős, P. and G. Szekeres, A combinatorial problem in geometry, *Comp. Math.* **2** (1935) 463–470.

100. A Permutation Game

A teacher sticks a number on the forehead of each of her six students, making sure that no two numbers are the same. No student can get any information about his own number, but can see the other five numbers. How can the students best divide themselves into two groups, group A and group B, say, with one (either A or B) consisting of the students with the largest, third largest and fifth largest numbers, and the other group consisting of the remaining three students?

Solution. Let the students be numbered $1, 2, \ldots, n$, although in the problem we have $n = 6$. By sticking various numbers on their foreheads, the teacher defines a permutation π of these numbers, but each student knows only the part of the permutation without his own number. For student i, let π_i be this $(n-1)$-permutation of the $n-1$ numbers in $[n] \setminus \{i\}$. Thus, for $n = 6$ and $\pi = 512643$ we have $\pi_6 = 51243$ and $\pi_4 = 51263$.

Each student constructs an n-permutation by moving his own number into the first place, and then takes the sign of this permutation. Those that get sign $+1$ form group A and those with sign -1 form group B. To spell it out, for each i, let ρ_i be the n-permutation starting with i and continuing with π_i. (This is the same as moving i in π into the first place.) Then student #i goes into group A if $\mathrm{sgn}(\rho_i) = 1$, and into B if $\mathrm{sgn}(\rho_i) = -1$. Continuing the example above, $\rho_6 = 651243$ and $\mathrm{sgn}(\rho_6) = 1$, $\rho_4 = 451263$ and $\mathrm{sgn}(\rho_4) = -1$, so 6 goes into A and 4 into B. But why is this always the right partition?

Simply because if there are k terms before i in the original full permutation π then $\mathrm{sgn}(\rho_i) = (-1)^k \mathrm{sgn}(\pi)$, so the sequence

$$\mathrm{sgn}(\rho_{\pi(1)}), \ \mathrm{sgn}(\rho_{\pi(2)}), \ldots, \ \mathrm{sgn}(\rho_{\pi(n)})$$

is an alternating sequence of $+1$ and -1. $\qquad\square$

Note. Experience shows that this problem is too easy for mathematicians, and too hard for others.

101. Ants on a Rod

There are 50 *ants on a rod of length* 1 m. *Each ant is hurrying along at* 10 mm/sec *in a fixed direction. When two ants meet, they turn around and each hurries in the opposite direction with the same speed. An ant that reaches the end of the rod falls off. The maximum time it may take for the ants to fall off the rod is* 100 sec.

Proof. Let us reinterpret the movement of the ants. Suppose when two ants meet, instead of turning around, they just continue their journeys. (As most of us are not too good at distinguishing one ant from another, this assumption is not unreasonable.) This does not change the dynamics of the system. Hence the longest it may take for our rod to be ant-free is the longest a single ant may spend on the rod. Running from one end to the other, an ant will take $1000/10 = 100$ sec. □

102. Two Cyclists and a Swallow

Two cyclists, A and B, start from a distance of 60 km towards each other, with A moving at 20 km/h and B at 10 km/h. A swallow keeps flying at 40 km/h from one to the other, starting with the faster cyclist, turning around as soon as it reaches the cyclist coming in the opposite direction. Show that the swallow will cover 80 km by the time the two cyclists meet.

Solution. The cyclists will meet in 60/(20+10)=2 hours; during that time the swallow will cover $2 \cdot 40 = 80$ km. □

Notes. My interest in this old chestnut that could be found in a collection of puzzles 60 years ago is its connection to John von Neumann, the great mathematician, illustrating his speed of thought. According to legend, when von Neumann was given this problem at one of the lavish parties he gave in his house in Princeton, he replied without hesitation, giving the correct answer. 'Ah, so you have found the trick', acknowledged the questioner. 'What trick?', retorted von Neumann, 'I have worked out the geometric series'.

What geometric series did von Neumann mean? The one that arises if we work out the answer in a plodding way. It takes the swallow $60/(40+10) = 6/5$ hours and 48 km to reach B; at that time the two cyclists are at $60 - \frac{6}{5}(20 + 10) = 24$ km from each other. Having turned around, the swallow will return to B in $24/(40 + 20) = \frac{2}{5}$ hours, covering 16 km in the process. At that time the two cyclists will be $24 - \frac{2}{5}(20 + 10) = 12$ km from each other, with the swallow next to B. Thus, after the swallow has flown $48 + 16 = 64$ km, we are back in the starting setup, except the two cyclists are only $12 = 60/5$ km apart, so every to and fro takes 1/5 times as long as the previous. Altogether, the swallow will cover

$$64\left(1 + 1/5 + (1/5)^2 + (1/5)^3 + \cdots\right) = 64\frac{1}{1 - 1/5} = 80 \text{ km},$$

as we know.

I heard this story about von Neumann in Budapest, when I was very young, and swallowed it hook, line and sinker. Shortly after I turned 20 and got to Cambridge, I asked the great physicist Paul Dirac about this story, and was surprised to hear that it may well have been true.

103. Almost Disjoint Subsets of Natural Numbers

There is a continuum family $\{M_\gamma : \gamma \in \Gamma\}$ of sets of natural numbers such that if $\alpha, \beta \in \Gamma$, $\alpha \neq \beta$, then $M_\alpha \cap M_\beta$ is a finite initial segment of both M_α and M_β.

First Proof   Let Γ be the set of all infinite 0–1 sequences $\gamma = \gamma_0 \gamma_1 \cdots$ with $\gamma_0 = 1$: this will be our index set. Clearly, Γ is not only uncountable, but of continuum cardinality. For $\gamma \in \Gamma$ set

$$M_\gamma = \left\{ \sum_{i=0}^{n} \gamma_i \, 2^{n-i} : \ n = 0, 1, \ldots \right\}.$$

Thus, e.g. if $\gamma = 110100 \cdots$ then $M_\gamma = \{1, 3, 6, 13, 26, 52, \ldots\}$.

To complete the proof, we must check that the continuum family $\{M_\gamma : \gamma \in \Gamma\}$ has the required properties. Given $\alpha, \beta \in \Gamma$, $\alpha \neq \beta$, if $N = \max\{i : \alpha_i = \beta_i\}$, then

$$M_\alpha \cap M_\beta = \left\{ \sum_{i=0}^{n} \gamma_i \, 2^{n-i} : n = 0, 1, \ldots, N \right\},$$

so $M_\alpha \cap M_\beta$ is indeed an initial segment of both M_α and M_β.   □

Second Proof   Let us write Γ for the open interval $(1, 2) \subset \mathbb{R}$ of real numbers between 1 and 2: this is our uncountable index set. As in the first proof, Γ is not only uncountable but is of continuum cardinality. For $\gamma \in \Gamma$ define

$$M_\gamma = \{\lfloor 2^n \gamma \rfloor : n = 1, 2, \ldots \}.$$

Clearly, $2 \leq \gamma(1) < \gamma(2) < \cdots$, so each M_γ can be viewed as an infinite sequence.

To complete the proof, we have to check that if $\alpha, \beta \in \Gamma$ and $\alpha \neq \beta$, then $M_\alpha \cap M_\beta$ is a finite initial segment of both M_α and M_β. To this end, first suppose that $\lfloor 2^m \alpha \rfloor = \lfloor 2^m \beta \rfloor$. Since $2^m \leq \lfloor 2^m \alpha \rfloor < 2^{m+1}$ and $2^n \leq \lfloor 2^n \beta \rfloor < 2^{n+1}$, this tells us that $m = n$. Also, if $\lfloor 2^n \alpha \rfloor \neq \lfloor 2^n \beta \rfloor$, say, $2^n \alpha < N = \lfloor 2^n \beta \rfloor$, then

$$\lfloor 2^m \alpha \rfloor < 2^{m-n} N \leq \lfloor 2^m \beta \rfloor$$

for every $m \geq n$. Hence, $M_\alpha \cap M_\beta$ is indeed a finite initial segment of both M_α and M_β. $\qquad\square$

Notes. I first heard of this problem from Alfréd Rényi in 1964 at Cambridge. When he challenged me with it, I came up with the first proof above, and then he showed me the elegant second proof. As Rényi informed me, the assertion is a special case of a result of Alfred Tarski.

In the 1970s I often put this problem on my *For the Enthusiast* examples sheets, when supervising first-year mathematics undergraduates in Trinity College, Cambridge. It was considered to be harder than Problem 10 in *Coffee Time in Memphis*, asking only for an uncountable set of subsets of the natural numbers such that any two intersect in only finitely many numbers. According to Paul Erdős, this latter assertion was folklore in the first half of the 20th century, with the standard example $M_\gamma = \{2^n + \lfloor n^\gamma \rfloor : n = 1, 2, \ldots\}$, where γ is a positive real number.

Reference

Tarski, A., Sur la décomposition des ensembles en sous ensembles presque disjoint, *Fund. Math.* **14** (1929) 205–215.

104. Primitive Sequences

Given $b \geq 1$, the maximal value of n for which there is a primitive sequence $(a_i)_1^n$ with $0 < a_i \leq b$ for every i is $\lfloor (b+1)/2 \rfloor$.

Proof. (i) Let $0 < a_1 < a_2 < \cdots < a_n \leq b$ be a primitive sequence. We shall show that $n \leq \lfloor (b+1)/2 \rfloor$.

We start with an observation. Let $\ell \geq m$ be non-negative integers, and $1 \leq i, j \leq n, i \neq j$. Then

$$|2^\ell a_i - 2^m a_j| \geq |2^{\ell-m} a_i - a_j| \geq 1. \tag{1}$$

Turning to our problem, for $1 \leq i \leq n$, let k_i be the maximal integer such that $2^{k_i} a_i \leq b$, and set $b_i = 2^{k_i} a_i$. Then $b/2 < b_i \leq b$ for every i and, by (1), we have

$$|b_i - b_j| \geq 1$$

whenever $i \neq j$. Hence, with $b_{\max} = \max_{1 \leq i \leq n} b_i$ and $b_{\min} = \min_{1 \leq i \leq n} b_i$ we find that

$$b/2 < b_{\min} \leq b_{\max} - (n-1) \leq b_{\max} \leq b,$$

so, again by (1),

$$2b_{\min} \geq b_{\max} + 1.$$

Hence

$$2(b_{\max} - (n-1)) \geq b_{\max} + 1,$$

so

$$2n - 1 \leq b_{\max} \leq b,$$

telling us that $n \leq \lfloor (b+1)/2 \rfloor$.

(ii) To complete our proof, we have to show that n can be as large as claimed,

272

i.e. given $b \geq 1$, for $n = \lfloor (b + 1)/2 \rfloor$ there is a primitive sequence $(a_i)_1^n$ with $0 < a_i \leq b$ for every i.

To see this, for $i = 1, \ldots, n$, define $a_i = b - (n - i)$, so that $a_1 < \cdots < a_n = b$. Then $|a_i - a_j| \geq 1$ for $i \neq j$, and

$$2a_1 = 2(b - n + 1) = b + 1 + (b - 2n + 1) \geq b + 1 = b_n + 1,$$

so $(a_i)_1^n$ is a primitive sequence. □

Notes. This problem is an extension of the first part of Problem 2 in *Coffee Time in Memphis*. In fact, we have proved more than claimed: if a sequence $0 < a_1 < \cdots < a_n \leq b$ is such that $|2^k a_i - a_j| \geq 1$ whenever $i \neq j$ and k is a non-negative integer then $n \leq \lfloor (b + 1)/2 \rfloor$. However, in this formulation this would be an easier problem.

105. The Time of Infection on a Grid

Starting with a set of n initially infected sites in G_n, the minimal time to full infection is $n - 1$.

Proof. That full infection can happen after $n - 1$ steps is trivial: start with the n sites of the main diagonal infected. Then at time one the sites on the three longest diagonals are infected, at time two the sites on the five longest, etc., at time $n - 1$ the sites on the $2n - 1$ longest diagonals parallel to the main diagonal. But at that stage all sites are infected.

Resembling Problem 34 in *Coffee Time in Memphis*, the only problem is that the infection cannot happen faster: no matter how we choose n sites to infect the entire grid, infection will take at least $n - 1$ steps.

Note that our grid G_n has n^2 sites and $2n(n - 1)$ edges; $(n - 2)^2$ of the sites have degree four, $4(n - 2)$ have degree three, and four, the four 'corners', have degree two. Thinking of cells instead of sites, we have n^2 cells, $(n - 2)^2$ of those have four neighbours, $4(n - 2)$ have three neighbours, and the four corner cells have two neighbours.

We know from Problem 34 of *Coffee Time in Memphis* that to infect the $n \times n$ board we need at least n initially infected cells. Even more, if n initially infected cells lead to full infection then

(i) every cell is infected by having precisely two infected neighbours,
(ii) of every two neighbouring cells, eventually one infects the other.

Let S_0 be a set of n initially infected cells in G_n that lead to full infection of G_n. Condition (ii) implies that no two cells in S_0 are neighbours. Also, if a cell is not in S_0, then two of its edges will be used to infect it, and through the remaining edges it will take part in infecting other cells. In particular, unless it is one of the four corner cells, it does infect another cell if it has three neighbours, and two other cells if it has four neighbours.

274

Let us number the cells at the times they get infected, with their infection time. Thus the cells in S_0 get 0, and a cell infected by two cells numbered u and w gets numbered $\max\{u, w\} + 1$. Our task is to show that some cell will be numbered $n - 1$.

Let us reiterate what we know about this numbering. Every cell numbered $t \geq 1$ has precisely two neighbours numbered at most $t - 1$, with at least one numbered precisely $t - 1$, and every other neighbour is numbered at least $t + 1$. In particular, for every cell x_1 numbered t_1 there is a path $x_1 x_2 \cdots x_\ell$ such that x_ℓ is one of the corner cells and, for $i \geq 2$, the cell x_i is numbered t_i, where $t_i > t_{i-1}$. Thus $t_\ell \geq t_1 + \ell - 1$.

This preparation is more than enough to prove our assertion. Let us write $t(x)$ for the number (label) of a cell x, so that if $x \in S_0$ then $t(x) = 0$. We distinguish two cases according to the parity of n. First, if n is odd, say, $n = 2k + 1$, then the grid G_n has a central cell z_1 at distance $2k = n - 1$ from *each* of the four corner cells, as in Figure 47. Hence if $z_1 z_2 \cdots z_\ell$ is a path such that z_ℓ is a corner cell and $t(z_1) < t(z_2) < \cdots < t(z_\ell)$ then $\ell \geq n$ and so

$$t(z_\ell) \geq t(z_1) + (\ell - 1) \geq n - 1.$$

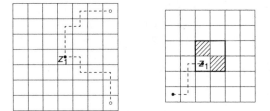

Figure 47   The cases n odd and n even.

Second, if n is even, say, $n = 2k$, consider the central 2×2 square of the grid, made up of four 'neighbouring' cells, as in Figure 47. At least one of these four cells, z_1, say, is not initially infected, so its infection time is at least 1: $t(z_1) \geq 1$. We know that there is a path $z_1 z_2 \cdots z_\ell$ from z_1 to a corner cell z_ℓ such that $t(z_1) < t(z_2) < \cdots < t(z_\ell)$. Since $\ell \geq 2k - 1 = n - 1$, we have

$$t(z_\ell) \geq t(z_1) + (\ell - 1) \geq 1 + (n - 2) = n - 1,$$

completing our proof.   □

106. Areas of Triangles: Routh's Theorem

(i) *Given a triangle ABC, let D, E and F be points on the sides BC, CA and AB, respectively. Then the segments AD, BE and CF are Cevians (i.e. concurrent) if and only if*

$$\frac{AF}{FB} \cdot \frac{BD}{DC} \cdot \frac{CE}{EA} = 1. \tag{1}$$

(ii) *A line ℓ meets the sides AB and BC of a triangle ABC in D and E, and the extension of the side AC in F. Then*

$$\frac{AD}{DB} \cdot \frac{BE}{EC} \cdot \frac{CF}{FA} = 1. \tag{2}$$

(iii) *On the sides BC, CA, AB of a triangle three points, A', B', C', are taken such that*

$$BA' : A'C = p_1 : q_1, \qquad CB' : B'A = p_2 : q_2, \qquad AC' : C'B = p_3 : q_3.$$

The segments BB' and CC', CC' and AA', AA' and BB' intersect in A'', B'', C''. Then the area of the triangle A''B''C'' is to the area of the triangle ABC as

$$(p_1p_2p_3 - q_1q_2q_3)^2 : (p_2p_3 + q_2q_3 + p_2q_3)(p_3p_1 + q_3q_1 + p_3q_1)(p_1p_2 + q_1q_2 + p_1q_2).$$

Proof. (i) Suppose that AD, BE and CF are Cevians, and denote their common intersection point by O. Take a line through B parallel to the side AC, and write K and L for its intersection points with the lines AD and CF, as in Figure 48. To deduce (1), we note two pairs of similar triangles and two similar degenerate quadrilaterals:

$$AFC \approx BFL, \qquad BDK \approx CDA \qquad \text{and} \qquad CEAO \approx LBKO.$$

276

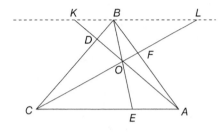

Figure 48 The Cevians AD, BE and CF meeting in O.

Indeed, these similarities imply

$$\frac{AF}{FB} = \frac{AC}{BL}, \qquad \frac{BD}{DC} = \frac{BK}{AC} \qquad \text{and} \qquad \frac{CE}{EA} = \frac{BL}{BK}.$$

Multiplying these three identities, (1) follows.

Conversely, suppose (1) holds. Let AD and BE meet in O, and let CO meet AB in F', so that AD, BE and CF' are Cevians. Then, by what we have just shown, $AF'/F'B = AF/FB$, so $F = F'$, i.e. the segment CF also contains O.

(ii) Drop perpendiculars on ℓ from the vertices A, B, C of the triangle, and write A', B' and C' for the feet of these perpendiculars, as in Figure 49. From the three pairs of similar triangles

$$ADA' \approx BDB', \qquad BEB' \approx CEC', \qquad CFC' \approx AFA'$$

we deduce that

$$\frac{AD}{DB} = \frac{AA'}{BB'}, \qquad \frac{BE}{EC} = \frac{BB'}{CC'}, \qquad \frac{CF}{FA} = \frac{CC'}{AA'}.$$

Multiplying these three identities, we obtain (2).

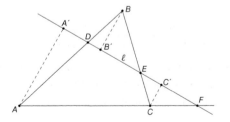

Figure 49 The points D, E, F, A', B' and C' on the line ℓ.

(iii) Applying (2) to the triangle ABB' and the line through C, A'', B'' and C', as in Figure 50, we find that

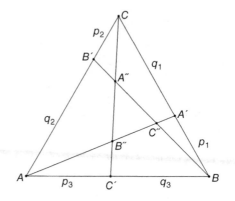

Figure 50 Constructing the triangle $A''B''C''$ from ABC.

$$\frac{AC'}{C'B} \cdot \frac{BA''}{A''B'} \cdot \frac{B'C}{CA} = 1,$$

so

$$\frac{BA''}{A''B'} = \frac{AC}{B'C} \cdot \frac{BC'}{C'A} = \frac{p_2 + q_2}{p_2} \cdot \frac{q_3}{p_3}.$$

Consequently,

$$\frac{\text{area}(BA''C)}{\text{area}(BB'C)} = \frac{BA''}{BB'} = \frac{p_2 + q_2}{p_2} \cdot \frac{q_3}{p_3} \bigg/ \left(\frac{p_2 + q_2}{p_2} \cdot \frac{q_3}{p_3} + 1\right)$$

$$= \frac{(p_2 + q_2)q_3}{p_2 p_3 + q_2 q_3 + p_2 q_3}.$$

Also, trivially,

$$\frac{\text{area}(BB'C)}{\text{area}(BAC)} = \frac{B'C}{AC} = \frac{p_2}{p_2 + q_2},$$

so

$$\frac{\text{area}(BA''C)}{\text{area}(BAC)} = \frac{p_2 q_3}{p_2 p_3 + q_2 q_3 + p_2 q_3}.$$

We obtain analogous expressions for the areas of the triangles $CB''A$ and $AC''B$. Finally, as ABC is the union of the four triangles $BA''C$, $CB''A$, $AC''B$ and $A''B''C''$ with pairwise disjoint interiors, we have

$$\text{area}(A''B''C'') = 1 - \frac{p_2 q_3}{p_2 p_3 + q_2 q_3 + p_2 q_3} - \frac{p_3 q_1}{p_3 p_1 + q_3 q_1 + p_3 q_1}$$

$$- \frac{p_1 q_2}{p_1 p_2 + q_1 q_2 + p_1 q_2}.$$

After some reduction, this tells us that if ABC has unit area then the area of the triangle $A''B''C''$ is

$$\frac{(p_1p_2p_3 - q_1q_2q_3)^2}{(p_2p_3 + q_2q_3 + p_2q_3)(p_3p_1 + q_3q_1 + p_3q_1)(p_1p_2 + q_1q_2 + p_1q_2)},$$

as claimed.   □

Notes. The first two parts are among the best known results in elementary plane geometry: (i) is *Ceva's theorem*, proved by the Italian *Giovanni Ceva* in the second half of the 17th century, and (ii) is the theorem of the Greek *Menelaus of Alexandria* from the 1st century AD. Sadly, today these results are much less known than they were sixty or a hundred years ago – that is why I have dared to put them in this collection. In fact, the theorem of Menelaus is usually stated in its signed form: a fraction like AD/DB is with a plus sign if AD and DB point in the same direction, i.e. D is on the side AB, otherwise it gets a minus sign. Then the product of the three fractions in (2) is, trivially, -1.

Ceva's theorem has several immediate consequences, e.g. the three medians of a triangle are concurrent (meeting in the centroid), and so are the three segments joining each vertex to the point of contact of the incircle with the opposite side. The point in which these three segments are concurrent is the *Gergonne point* of the triangle.

We have stated (ii) and its analogue, (i), because we have used it to prove (iii), *Routh's theorem*. As only the ratios p_i/q_i matter, we may assume that $q_i = 1$. Then, taking 1 for the area of the triangle ABC, Routh's theorem says that the area of $A''B''C''$ is

$$\frac{(p_1p_2p_3 - 1)^2}{(p_1p_2 + p_1 + 1)(p_2p_3 + p_2 + 1)(p_3p_1 + p_3 + 1)}.$$

Note that this result contains (i), Ceva's theorem: the area of the triangle $A''B''C''$ is zero if and only if the segments AA', BB' and CC' are concurrent.

Routh's theorem has a strange history. It first appeared in print in 1891 in the first edition of a book of Routh, and then in its second edition in 1896. The result is only stated in a footnote, together with the trivial observation that, in the notation of Figure 50, and taking signed areas in the obvious sense, depending on the orientation,

$$\frac{\text{area}(A''B''C'')}{\text{area}(ABC)} = \frac{p_1p_2p_3 - q_1q_2q_3}{(p_1 + q_1)(p_2 + q_2)(p_3 + q_3)}.$$

Routh adds that he has not met with these expressions before. That is *most unlikely*, because part (iii), i.e. Routh's theorem, appeared in the Tripos exam in 1878, set by Glaisher. As Routh was the most eminent of the Cambridge

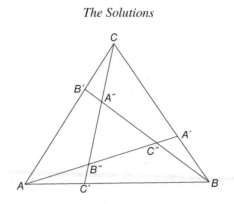

Figure 51 One third of the sides to one seventh of the area.

coaches ever, training every one of the Senior Wranglers in 22 consecutive years, it is most likely that he had set that problem to his pupils quite a few times. My explanation, based on no evidence, is that Routh had discovered his theorem before 1878 and, with his consent, the junior Glaisher borrowed it from him as a Tripos question.

In the fierce debate leading to the abolishment of the strict order of merit (and so the distinction of being Senior Wrangler) in the Mathematical Tripos after the 1909 exam, Edward Routh, himself a Senior Wrangler in 1854, fought a losing battle against three Fellows of Trinity College, James Glaisher (2W 1871), Andrew Forsyth (SW 1881) and the great G.H. Hardy (4W 1898).

As the final word, let us add that Routh's theorem often occurs in the special form when A', B', C' cut off one third of the sides, and so the area of $A''B''C''$ is one seventh of the area of ABC, as in Figure 51 and in the lovely book by Steinhaus.

References

Routh, E.J., *Treatise on Analytical Statics with Numerous Examples*, Vol. 1, Second edition, Cambridge University Press (1896) (see p. 82).

Glaisher, J.W.L. (ed.), *Solutions of the Cambridge Senate-House Problems and Riders for the Year 1878*, Macmillan (1879).

Steinhaus, H., *Mathematical Snapshots*. New edition, revised and enlarged, Oxford University Press (1960).

107. Lines and Vectors – Euler and Sylvester

(i) *Let O be the circumcentre of a triangle ABC, M the intersection of the medians and H the orthocentre. Then O, M and H are collinear. Furthermore OH = 3OM.*
(ii) *The resultant of the vectors $\overrightarrow{OA}$, $\overrightarrow{OB}$ and $\overrightarrow{OC}$ is $\overrightarrow{OH}$.*

Proof. (i) If the triangle *ABC* is right-angled or regular, the assertions are trivial. Hence, we may assume that $AC \neq BC$, $C \neq H$ and *O* is not C', the midpoint of *AB*. We know that *M* is on CC' and $CM = 2MC'$. Write *G* for

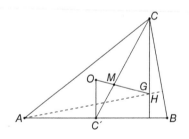

Figure 52 Our notation.

the point on the half-line *OM* satisfying $MG = 2OM$ (i.e. $OG = 3OM$), as in Figure 52. Then the triangles *CGM* and $C'OM$ are similar (with factor 2). Consequently the segments OC' and *GC* are parallel, so *CG* is orthogonal to *AB*, implying that *G* is on the altitude of *ABC* from *C*. Analogously, *G* is on every altitude, so $G = H$, completing our proof that *O*, *M* and *H* are collinear and $OH = 3OM$.

(ii) Since C' is the midpoint of *AB*, and *M* is the third from C' to *C*,

$$\overrightarrow{OC'} = \frac{1}{2}\overrightarrow{OA} + \frac{1}{2}\overrightarrow{OB} \qquad \text{and} \qquad \overrightarrow{OM} = \frac{2}{3}\overrightarrow{OC'} + \frac{1}{3}\overrightarrow{OC},$$

282

we have

$$\overrightarrow{OH} = 3\overrightarrow{OM} = 2\overrightarrow{OC'} + \overrightarrow{OC} = \overrightarrow{OA} + \overrightarrow{OB} + \overrightarrow{OC},$$

as claimed.   □

Notes. Part (i) is due to Euler: the line through O, M and H is the *Euler line* of the triangle; part (ii) was first noted by Sylvester.

108. Feuerbach's Remarkable Circle

Let H be the intersection of the altitudes AD, BE and CF of a triangle ABC. Then the midpoints of the sides, the midpoints of the segments AH, BH, CH, and the feet D, E, F of the altitudes are on a circle.

Proof. Let A', B', C' be the midpoints of the sides BC, CA and AB, respectively. Then $B'C' = \frac{1}{2}BC$; furthermore, as the angle BFC is $\pi/2$ and A' is the midpoint of BC, the segment $A'F$ is also of length $\frac{1}{2}BC$ (see the first triangle in Figure 53). Hence the trapezoid $A'B'C'F$ is isosceles, so F is on the circle through $A'B'C'$. Analogously, D and E are also on this circle. Turning this around, the circle through D, E and F goes through the midpoints of the sides.

What does this say about the triangle ABH instead of ABC? Amazingly, in this triangle the feet of the altitudes are exactly as before, in ABC, namely D, E and F, only this time E belongs to A and D to B, as shown in the second triangle in Figure 53. Consequently, the circle through D, E and F goes through the

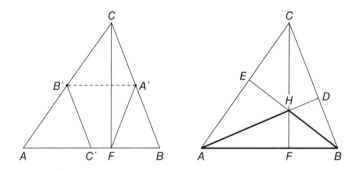

Figure 53 On the left is shown the trapezoid $A'B'C'F$, and on the right is the triangle ABH, with D, E and F the feet of its altitudes.

midpoints of the sides *AH* and *BH*. Analogously, it goes through the midpoint of *CH* as well, completing our proof. □

Notes. This remarkable circle was known to Leonhard Euler in 1765, but it is called the *Feuerbach circle* after Karl Feuerbach, who rediscovered it in 1822. As the circle goes through the midpoints of the sides and the feet of the altitudes, its centre is on the Euler line (see Problem 107), halfway between the orthocentre, *H*, and the circumcentre, *O*.

109. Euler's Ratio–Product–Sum Theorem

Let ABC be a triangle, and x, y, z positive numbers. Then the following two assertions are equivalent.

(i) There are points X, Y and Z on the sides BC, CA and AB such that the segments AD, BY and CZ are concurrent in a point O, and $AO/OX = x$, $BO/OY = y$ and $CO/OZ = z$.
(ii) The numbers x, y and z satisfy

$$xyz = x + y + z + 2.$$

Proof. We may assume that area$(ABC) = 1$. Our assertion is an easy consequence of the following claim.

Claim. Given positive numbers x and y, there is a unique point O in the triangle ABC such that $AO/OX = x$ and $BO/OY = y$.

Indeed, if $AO/OX = x$ for a point O on O on AX, with X on BC, then area$(BOC) = OX/AX = 1/(1+x)$. Similarly, if $BO/OY = y$ then area$(COA) = 1/(1 + y)$. The locus of points P with area$(BPC) = 1/(1 + x)$ is a line parallel to BC, and the locus of points Q with area$(CQA) = 1/(1 + y)$ is a line parallel to AC. These two lines intersect in a unique point O of the triangle, proving our Claim.

  Consequently, (i) holds if and only if the point O determined by X and Y satisfies area$(AOB) = 1/(1+z)$. Since the three triangles BOC, COA and AOB partition ABC, (i) holds if and only if

$$\frac{1}{1 + x} + \frac{1}{1 + y} + \frac{1}{1 + z} = 1,$$

i.e.

$$xyz = x + y + z + 2,$$

as required. □

Notes. The result above was proved by Leonhard Euler in 1780 by algebra and trigonometry. In 1999 Shephard gave a considerably simpler proof of the result that (i) implies (ii), i.e. if X, Y and Z are points on the sides BC, CA and AB of a triangle ABC such that the segments AX, BY and CZ are concurrent in a point O of ABC then

$$\frac{AO}{OX} \cdot \frac{BO}{OY} \cdot \frac{CO}{OZ} = \frac{AO}{OX} + \frac{BO}{OY} + \frac{CO}{OZ} + 2.$$

It is not surprising that there has been much recent work on extensions of Euler's ratio–product–sum theorem: for some of these extensions, see the references.

References

Euler, L., Geometrica et sphaerica quaedam, *Mem. Acad. Sci. St. Petersb.*, **5** (1815) 96–114; *Opera Omnia Series 1, vol. XXVI*, pp. 344–358. Original: http://eulerarchive.maa.org/docs/originals/E749.pdf.
English translation: http://eulerarchive.maa.org/Estudies/E749t.pdf.

Grünbaum, B., Cyclic ratio sums and products, *Crux Math.* **24** (1998) 20–25.

Grünbaum, B., and M.S. Klamkin, Euler's ratio–sum theorem and generalizations, *Math. Mag.* **79** (2006) 122–130.

Kozma, J., and Á. Kurusa, Hyperbolic is the only Hilbert geometry having circumcenter or orthocenter generally, *Beitr. Algebra Geom.* **57** (2016) 243–258.

Kurusa, Á., and J. Kozma, Euler's ratio–sum formula in projective-metric spaces, *Beit. Algebra Geom.* **60** (2019) 379–390.

Papadopoulos, A., and W. Su, On hyperbolic analogues of some classical theorems in spherical geometry. In *Hyperbolic Geometry and Geometric Group Theory*, Advanced Studies in Pure Mathematics **73**, Mathematical Society of Japan (2017) pp. 225–253.

Shephard, G.C., Euler's triangle theorem, *Crux Math.* **25** (1999) 148–153.

110. Bachet's Weight Problem

We have a weighing scale with two pans. Denote by W_n the maximal integer for which there is a set of n integral weights that can weigh any integral number of pounds by putting weights in either pan.

(i) *Then $W_n = (3^n - 1)/2$.*
(ii) *The only set of weights that will weigh any weight up to $(3^n - 1)/2$, when weights may be put in either pan, is $\{1, 3, 9, \ldots, 3^{n-1}\}$.*

Proof. If $w_1 \leq w_2 \leq \cdots \leq w_n$ are our weights then we can weigh an object of weight p if and only if there is a sequence $(\varepsilon_i)_1^n$, $\varepsilon_i = +1, -1$ or 0, such that

$$p = \sum_{i=1}^{n} \varepsilon_i \, w_i.$$

Indeed, this relation is a simple encoding of the position of the weights when p is weighed: $\varepsilon_i = -1$ if weight w_i is put into the same pan as p, $\varepsilon_i = +1$ if w_i is a counterweight, and $\varepsilon_i = 0$ if weight w_i is not used in weighing p.

(i) There are 3^n sequences $(\varepsilon_i)_1^n$. One of them is the all-zero sequence, not used to measure any weight p, $1 \leq p \leq W_n$. The other $3^n - 1$ sequences come in pairs, $(\varepsilon_i)_1^n \leftrightarrow (-\varepsilon_i)_1^n$, with one weighing a positive number p and the other $-p$, in the obvious sense. Hence, at most half of the sequences $(\varepsilon_i)_1^n$ can be used to weigh an object, so $W_n \leq (3^n - 1)/2$.

Also, with $w_i = 3^{i-1}$, $1 \leq i \leq n$, every integral weight $p \leq (3^n - 1)/2$ can be weighed. Indeed, every positive integer $m < 3^n$ has a ternary expansion of the form

$$m = \sum_{i=0}^{n-1} c_i \, 3^i,$$

where $c_i = 0, 1$ or 2. Hence, for $1 \leq p \leq \sum_{i=0}^{n-1} 3^i = (3^n - 1)/2$, we have

$$p + \sum_{i=0}^{n-1} 3^i = \sum_{i=0}^{n-1} c_i \, 3^i,$$

where $c_i = 0, 1$ or 2. Consequently,

$$p = \sum_{i=0}^{n-1} (c_i - 1) 3^i = \sum_{i=0}^{n-1} \varepsilon_i \, 3^i,$$

where $\varepsilon_i = c_i - 1 = 1, -1$ or 0. Thus $W_n = (3^n - 1)/2$, as claimed.

(ii) Let $1 \leq w_1 \leq \cdots \leq w_n$ be a sequence of n weights achieving $W_n = (3^n - 1)/2$. Then $\sum_1^n w_i = W_n = (3^n - 1)/2$, and, for $\varepsilon = +1, -1$ or 0, all expansions $\sum_1^n \varepsilon_i w_i$ (of positive or negative integers) are different. This implies that all w_i are different integers: $1 \leq w_1 < w_2 < \cdots < w_n$, and each w_i is an integer. Also, since the second largest weight our weights can weigh is $w_2 + w_3 + \cdots + w_n = W_n - 1$, we must have $w_1 = 1$. To complete our proof, we show by induction on k that $w_k = 3^{k-1}$ for every k, with $1 \leq k \leq n$.

Suppose that $w_1 = 1, w_2 = 3, \ldots, w_k = 3^{k-1}$. With these k weights, every weight up to $(3^k - 1)/2$ can be weighed, so for every $W_n - (3^k - 1) \leq p \leq W_n$ there is an expansion

$$p = \sum_{i=1}^{k} \varepsilon_i 3^{i-1} + \sum_{i=k+1}^{n} w_i.$$

These are the only expansions containing $w_{k+1} + w_{k+2} + \cdots + w_n$. The smallest of them is

$$- \sum_{i=1}^{k} 3^{i-1} + \sum_{i=k+1}^{n} w_i = W_n - 3^k + 1.$$

The expansion of the weight only one smaller than this must be

$$\sum_{i=1}^{k} 3^{i-1} + \sum_{i=k+2}^{n} w_i = W_n - 3^k,$$

so

$$w_{k+1} = 2 \sum_{i=1}^{k} 3^{i-1} + 1 = 3^k.$$

This completes our induction, so our result is proved. □

Notes. This problem is probably the best known weighing problem. It is usually attributed to Claude-Gaspar Bachet, who in 1612 proposed the question of weighing all objects up to 40 pounds: '*Estant proposée telle quantité qu'on voudra pesant un nombre de livres depuis* 1 *jusques à* 40 *inclusivement (sans toutesfois admettre les fractions) on demande combien de pois pour le moins il faudroit employer à cet effect*'. This is then the case $n = 4$ of the problem above.

Bachet's problem was taken up and generalized by Major MacMahon in a paper in 1886, and later in his famous book first published in 1915. As it so often happens, recently it was discovered that calling it Bachet's problem is a misnomer, because the problem is much older. Knobloch tells us that the problem for weight 40 was first solved by Fibonacci (Leonardo Pisano) in 1202. Indeed, Fibonacci, who was probably the greatest mathematician in the Middle Ages, considered the problem in his famous book, *Liber Abaci*.

Knobloch made a thorough study of the history of this problem: he listed twelve others who had solved the problem before Bachet and six others who solved it independently in the 17th century. Thus, as pointed out by Knobloch, partition theory did not begin with Euler's work on 'partitio numerorum'.

For further fascinating facts about this problem, see O'Shea's paper from 2010.

References

Bachet, C.-G., *Problèmes Plaisants et dÉlectables, qui se font par les Nombres*, 5ième éd. Revue, simplifiée et augmentée par A. Labosne, Librairie Scientifique et Technique Albert Blanchard (1959).

Knobloch, E., Zur Überlieferungsgeschichte des Bachetschen Gewichtsproblems, *Sudhoffs Arch.* **57** (1973) 142–151.

MacMahon, P.A., Certain special partitions of numbers, *Q. J. Maths* **21** (1886) 367–373.

O'Shea, E., Bachet's problem: As few weights to weigh them all, ArXiv:1010.5486, 2010, 15 pp.

Pisano, L., *Fibonacci's Liber Abaci*. A translation into modern English of Leonardo Pisano's *Book of Calculation*, translated from the Latin and with an introduction, notes and bibliography by L.E. Sigler. Sources and Studies in the History of Mathematics and Physical Sciences, Springer-Verlag (2002).

111. Perfect Partitions

We have a weighing scale with two pans. One of the pans is for our integral weights, the other is for the object we wish to weigh.

Under these conditions, given a prime p and integer $\alpha \geq 1$, there are $2^{\alpha-1}$ ways of partitioning $n = p^\alpha - 1$ lb into integral weights so as to be able to weigh, in only one manner, any weight of an integral number of pounds from 1 to n inclusive. (Weights of the same weight are considered to be identical.) In particular, for $n = 31$ there are 16 suitable partitions.

Proof. A *partition* of an integer n is a representation of it as a sum of natural numbers. For example, $31 = 8 + 8 + 8 + 2 + 2 + 2 + 1$ is a partition of 31. The order of the summands is irrelevant, so one usually starts with the largest summands. It is customary to abbreviate a partition by putting the multiplicity of a part into the exponent, so that this partition of 31 is written as $(8^3 2^3 1)$. If the order of the summands is taken into account then we talk of a *composition* of n. Thus $2, 1, 2$ and $1, 2, 2$ are different compositions of 5. Clearly, 5 has two different partitions and six different compositions into three numbers.

A partition $(n_1^{\alpha_1} \cdots n_r^{\alpha_r})$ of $n = \sum_1^r \alpha_i n_i$ is *perfect* if it satisfies the conditions of the problem, i.e. if using these summands as weights, it is possible to weigh in one and only one manner any weight of an integral number of pounds from 1 to n inclusive. That $(8^3 2^3 1)$ is a perfect partition of 31 is equivalent to the polynomial identity

$$(1 + X^8 + X^{16} + X^{24})(1 + X^2 + X^4 + X^6)(1 + X) = 1 + X + X^2 + \cdots + X^{31}.$$

In general, a partition $(n_1^{\alpha_1} \cdots n_r^{\alpha_r})$ is perfect if and only if

$$\prod_{i=1}^r \left(\sum_{k=0}^{\alpha_i} X^{kn_i} \right) = 1 + X + X^2 + \cdots + X^n.$$

291

Indeed, this identity holds if and only if for every power X^m, $1 \le m \le n$, there is a unique sequence $(k_i)_1^r$, with $0 \le k_i \le \alpha_i$, such that $m = k_1 n_1 + k_2 n_2 + \cdots + k_r n_r$.

In our case, when $n = p^\alpha - 1$, it can be shown that such factorizations arise from products of the form

$$\frac{1 - X^{p^\alpha}}{1 - X^{p^{\alpha_1}}} \cdot \frac{1 - X^{p^{\alpha_1}}}{1 - X^{p^{\alpha_2}}} \cdot \ldots \cdot \frac{1 - X^{p^{\alpha_s}}}{1 - X},$$

where $\alpha - \alpha_1, \alpha_1 - \alpha_2, \ldots, \alpha_{s-1} - \alpha_s, \alpha_s$ is a composition of α. Conversely, every composition of α gives a suitable factorization and so a perfect partition of $p^\alpha - 1$, namely

$$\left((p^{\alpha_1})^{p^{\alpha-\alpha_1}-1}, (p^{\alpha_2})^{p^{\alpha_1-\alpha_2}-1}, \ldots, 1^{p^{\alpha_s}-1} \right).$$

For example, the composition 2, 1, 2 of $\alpha = 5$ gives the perfect partition

$$(p^3)^{p^2-1} \, (p^2)^{p-1} \, 1^{p^2-1}$$

of $p^\alpha - 1$, and 1, 3, 1 gives

$$(p^4)^{p-1} \, p^{p^3-1} \, 1^{p-1}.$$

We have shown that the perfect partitions of $p^\alpha - 1$ are in one-to-one correspondence with the compositions of α. As our last step, we have to show that an integer α has $2^{\alpha-1}$ compositions. To see this, note that a composition of α corresponds to a set of 'division lines' in the sequence $12 \cdots \alpha$; e.g. division lines after 2, 3, 5 and 8 for $\alpha = 9$, i.e. $12|3|45|678|9$, give the composition $2, 1, 2, 3, 1$ of 9. Hence, α has $2^{\alpha-1}$ compositions, as claimed, completing our proof.   □

Notes. The results above are due to Major Percy Alexander MacMahon (1854–1929), and are from the first volume of his two-volume treatise on 'Combinatory Analysis', pp. 217–223. Major MacMahon (as he was always called in his later years) is not as well known as he should be, considering that he was an important (should one say major?) mathematician, who did much work on partitions and wrote the first books on combinatorics in England. Ever since I first heard about him from Littlewood (to my shame, I had not heard of him before then), I have thought very highly of him. He was just about the most courteous mathematician in Cambridge, especially to young people, whom he was always eager to help, and was a great supporter of Srinivasa Ramanujan. It is widely known that Ramanujan was promoted by G.H. Hardy and J.E. Littlewood, but it is much less known that MacMahon also did everything to help Ramanujan. But Hardy had brought the brilliant Ramanujan to Trinity College, where Hardy and Littlewood were Fellows, and always considered him his discovery, while

Major MacMahon was at St. John's College. In his book, written soon after Ramanujan's arrival in Cambridge, he paid tribute to Ramanujan by having an entire chapter on 'Ramanujan's Identities'.

That Hardy and Ramanujan did some of their best work on the partition function $p(n)$, the number of partitions of n, probably had much to do with MacMahon, who had worked on $p(n)$ for years before them. (To be on the safe side, note that $p(4) = 5$, as the partitions of 4 are $4, 3 + 1, 2 + 2, 2 + 1 + 1$ and $1 + 1 + 1 + 1$.) MacMahon's numerical work on $p(n)$ inspired Hardy and Ramanujan to get a really sharp asymptotic expansion for $p(n)$. The first six terms of this asymptotic formula gave that

$$p(200) = 3,972,999,029,388.004;$$

to check the accuracy of this expansion, MacMahon's subsequent hand calculations gave

$$p(200) = 3,972,999,029,388,$$

an extraordinarily close fit.

Another reason why I have always felt for Major MacMahon is that he suffered a serious injustice. Upon the death of J.J. Sylvester in March 1897, the Savilian Chair of Geometry at Oxford became vacant. MacMahon's application for the chair was unsuccessful, and a blatantly unfair appointment was made. Although he received several honorary doctorates, including one from Cambridge, this sense of injustice remained with him all his life, much like the sense of misfortune that befell Paul Erdős.

Reference

MacMahon, P.A., *Combinatory Analysis*, Vols. I and II. Originally published in two volumes (1915, 1916) by Cambridge University Press. Reprinted as one volume, 1960, Chelsea Publishing Company (see pp. 217–223 of Vol. I).

112. Countably Many Players

Countably many players, $P_1, P_2, \ldots$, are lined up in a row; each of them has a real number on his head, which can be seen only by the players preceding him. Thus player P_3 has no idea what numbers are on P_1, P_2 and P_3 (himself), but can see the numbers on P_4, P_5, etc. The task of the players is to guess their own number, and write it on a piece of paper that no other player will see. Before the numbers are distributed, the players are allowed to agree on a strategy, but once the numbers have been placed on the hats, no communication is permitted. With a suitable strategy, the players can achieve that all but finitely many of them guess their numbers correctly.

Solution. Call two sequences of reals *equivalent* if they agree in all but finitely many terms. Thus $a = (a_n)_1^\infty \equiv b = (b_n)_1^\infty$ if there is a threshold number $n_0 = n_0(a, b)$ such that $a_n = b_n$ whenever $n \geq n_0$. This is clearly an equivalence relation, so every sequence a determines an equivalence class E_a. The four relations $a \equiv b$, $a \in E_b$, $b \in E_a$ and $E_a = E_b$ are clearly equivalent(!).

Before the game starts, the players pick a member of each equivalence class, and call it the '*canonical representative*' of that class. Any member of the class will do, the only condition is that every player should know which member of a class is its canonical representative. Thus, if E is an equivalence class then the players decide to represent it by a certain sequence $z_E \in E$ depending only on E.

Turning to the game, let $x = (x_n)_1^\infty$ be the sequence of numbers on the hats. The nth player knows all but n terms of this sequence, so he knows its equivalence class $E = E_x$, and therefore the canonical representative $z = z_E$ of this class. Player n simply guesses z_n for the number on his own hat: once $n \geq n_0(x, z)$, this guess is correct. $\qquad\square$

Notes. This problem tends to be the standard introduction to 'infinitely many hats' problems. It certainly demonstrates the power of the canonical representative of an equivalence class.

113. One Hundred Players

As usual in 'games' of this kind, one hundred players are allowed to agree on a strategy before the game starts, but after the start no consultation is permitted. One by one, the players are shown into a room with an infinite sequence of drawers, $D_1, D_2, \ldots$, with drawer D_n containing a real number x_n not known to any of the players. Each player is allowed to open as many drawers as he likes, but at some stage he has to point at an unopened drawer, and guess the real number in it. After this, the drawers are closed again, and the next player enters the room.

Remarkably, the players can guarantee that at least one player does guess correctly. In fact, even more remarkably, they can guarantee that at least 99 of the 100 players guess correctly.

Solution. As in the previous problem, the players, $P_1, \ldots, P_{100}$, agree on a canonical representative z_E of each equivalence class E of sequences of reals, i.e. for each equivalence class they pick a particular member of it. The players partition the set $\mathbb{N}$ into 100 infinite sequences, e.g. by taking

$$N_i = \{n \in \mathbb{N} : n \equiv i \bmod 100\} = \{i + 100(n-1) : n = 1, 2, \ldots\},$$

$i = 1, \ldots, 100$. A sequence of reals, $a = (a_n)_1^\infty$, is made up of sequences $a^{(i)}$ indexed with N_i, $i = 1, \ldots, 100$; indeed, $a^{(i)} = (a_n^{(i)})_{n=1}^\infty$, where $a_n^{(i)} = a_{i+100(n-1)}$. Note that a sequence $b = (b_n)_1^\infty$ is equivalent to a if and only if $a^{(i)}$ is equivalent to $b^{(i)}$ for every i.

Here is then how the players proceed. Let $x = (x_n)_1^\infty$ be the sequence of reals formed by the numbers in the drawers, and let z be the canonical representative of the equivalence class E_x of x. Of course, when the game starts, no player has any idea as to what z is. On entering the room, player P_i opens the drawers indexed by N_j, $j \neq i$, and thereby identifies 99 sequences $x^{(j)}$, $j \neq i$, their equivalence classes $E_{x^{(j)}}$, and the canonical representatives of these classes,

$z^{(j)}$, $j \neq i$. Furthermore, P_i identifies the thresholds t_j beyond which $x^{(j)}$ and $z^{(j)}$ agree, and sets $T_i = \max_{j \neq i} t_j$. Note that none of these quantities depends on i: the only condition is that $j \neq i$.

After this, P_i turns his attention to the drawers whose indices are in N_i, and opens the drawers indexed by the subset $\{i + 100n : n > T_i\}$ of N_i. At this stage, P_i knows all but a finite number of terms of x, so he knows the canonical representative z of the equivalence class E_x of x. Finally, P_i guesses that the drawer D_{i+100T_i} contains the number z_{i+100T_i}. When is this guess correct? It is certainly correct if $z^{(i)}$ and $x^{(i)}$ agree T_i onwards, which happens when

$$t_i \leq T_i = \max_{j \neq i} t_j.$$

This certainly holds if there exists some $j \neq i$ with $t_j \geq t_i$. So in conclusion we see that if the maximum value $\max t_i$ is attained at one t_i then every player except maybe P_i guesses correctly, while if this maximum is attained at more than one t_i then every player guesses correctly. □

Notes. I was reminded of this problem by Imre Leader, but I do not know whom one should credit with coming up with the problem. I first heard of it from Paul Erdős, who loved arguments based on equivalence classes, but he was not the one who invented the problem. The proof above is somewhat formal, perhaps too formal: the reader may amuse himself by rewriting it in a chattier style. I understand from Imre Leader that he has used this problem in his supervisions at Trinity College, and most of his students found it difficult.

114. River Crossings: Alcuin of York – Take One

(i) De viro et muliere ponderantibus plaustrum *[A very heavy man and woman.]*
A man and a woman, each the weight of a cartload, with two children who together weigh as much as a cartload, have to cross a river. They find a boat which can only take one cartload. Although it seems unlikely, with a suitable arrangement they can cross the river without sinking the boat.
(ii) De lupo et capra et fasciculo cauli *[A wolf, a goat and a bunch of cabbages.]*
A man had to take a wolf, a goat and a bunch of cabbages across a river. The only boat he could find could only take two of them at a time. By rowing back and forth quite a few times, the man managed to transfer all of these to the other side in good condition.

Solution. The beauty of these problems is not the elegance or difficulty of the mathematics involved, but their age; both are from the collection of puzzles Alcuin of York wrote in 799 or so to 'sharpen the minds of the young'. In each case, first we shall give Alcuin's solutions from his collection, in the translation of John Hadley, published by Singmaster and Hadley in 1992, and then we shall make some comments on the 'mathematics' involved.

(i) *Alcuin's Solution.* First, the two children get into the boat and cross the river; one of them brings the boat back. The mother crosses in the boat; and her child brings the boat back. His sister joins him in the boat and they go across; and again one of them brings the boat back to his father. The father crosses, and his son, who had previously crossed, having boarded, returns to his sister; and both cross again. With such ingenious navigation the crossing may be completed without shipwreck.

This problem is by far the easiest of Alcuin's 'river crossing' problems. As the parents have only one way to cross, since they have to be in the boat by themselves, the role of the children is simply to bring the boat to the starting bank, without getting stranded. A trivial arrangement is described by Alcuin above; in fact, this is the only solution without wasteful moves.

(ii) *Alcuin's Solution.* I would take the goat and leave the wolf and the cabbage. Then I would return and take the wolf across. Having put the wolf on the other side, I would take the goat back over. Having left that behind, I would take the cabbage across. I would then row across again, and having picked up the goat take it over once more. By this procedure there would be some healthy rowing, but no lacerating catastrophe.

This 'wolf–goat–cabbages problem' is less trivial than the 'fat people problem' in part (i): at least one cannot be expected to give the solution immediately. Still after a moment's thought one cannot go wrong, especially if one realizes that all the conditions mean is that one has to take care of the goat. The goat is either by itself, or its party contains the man. Thus, thinking of trips there and back, except the last time: 1. The man takes over the goat; 2. He takes over the wolf and brings back the goat; 3. He takes over the cabbages; 4. He takes over the goat and does not return.   □

Notes. As we have already remarked, the two problems above are from Alcuin's collection of mathematical puzzles, the oldest mathematical problem collection in Latin. W.W. Rouse Ball wrote in his *Short Account of the History of Mathematics*: 'This collection ... was the work of a man of exceptional genius ...'. Problems of the type proposed by Alcuin were taken up in the Middle Ages by Tartaglia and others, and revived again towards the end of the 19th century. For more recent papers on this topic see the references below.

Alcuin of York (c. 732–804) was a major figure in the Middle Ages: a clergyman, a mathematician, an astronomer, a poet, a biblical scholar, and a general educator. He was widely considered to be the most learned man anywhere to be found: he wrote texts on arithmetic, geometry and astronomy, greatly aiding the renaissance in learning in Europe.

Alcuin was a child when he joined the York Minster community; later he taught there and eventually became the head of the Cathedral School, which he turned into a centre of theology, liberal arts, literature and science. He revived the school with the trivium and quadrivium, and wrote a codex on the trivium.

He was sent to Rome on behalf of the King of Northumbria and on the way back he met Charlemagne (749–814) in Parma, who invited him to his court. Alcuin spent most of the 780s and 790s in Aachen, teaching Charlemagne himself, his sons Pepin and Louis, and many others, and became Charlemagne's friend and counsellor. He improved the curriculum in the Palace School, raised the standards of scholarship and encouraged the study of liberal arts. He also started to perfect the Carolingian minuscule, the forerunner of today's Roman font.

Interrupting his stay in Aachen for a year, he returned to England and re-organized the studies at his old school in York. In 801, he begged permission to retire from court: Charlemagne appointed him abbot of St Martin at Tours, asking him to be available if he needed his counsel. At Tours, Alcuin and his monks continued his work on the Carolingian minuscule script. Towards the end of his life, he wrote touchingly about his achievements: '*In the morning, at the height of my powers, I sowed the seed in Britain, now in the evening when my blood is growing cold I am still sowing in France, hoping both will grow, by the grace of God, giving some the honey of the holy scriptures, making others drunk on the old wine of ancient learning*'.

And in his epitaph he added:

Dust, worms, and ashes now ...
Alcuin my name, wisdom I always loved,
Pray, reader, for my soul.

References

Alcuin of York, Propositiones ad acuendos juvenes (Problems to sharpen the young). An annotated translation of the oldest mathematical problem collection in Latin, translated by J. Hadley, commentary by D. Singmaster and J. Hadley, *Math. Gaz.* **76** (1992) 102–126.

Ball, W.W. Rouse, *A Short Account of the History of Mathematics*, Macmillan (1888).

Fraley, R., K.L. Cooke and P. Detrick, Graphical solution of difficult crossing puzzles, *Math. Mag.* **39** (1966) 151–157.

Pressman, I. and D. Singmaster, The jealous husbands and the missionaries and cannibals, *Math. Gaz.* **73** (1989) 73–81.

Schwartz, B.L., An analytic method for the "difficult crossing" puzzles. *Math. Mag.* **34** (1960/61) 187–193.

115. River Crossings: Alcuin of York – Take Two

De tribus fratribus singulas habentibus sorores *[Three friends and their sisters.]* *Three men, each with a sister, needed to cross a river. Each one of them coveted the sister of another. At the river, they found only a small boat, in which only two of them could cross at a time. Using a suitable strategy, the party could cross the river, without any of the women being defiled by the men.*

Solution. As we have already said, the beauty of the river crossing problems is not their elegance nor is it their difficulty, but their provenance, the collection of puzzles Alcuin of York wrote in 799 or so to 'sharpen the minds of the young'. The present problem is the first of the four 'river-crossing' puzzles Alcuin gives – somewhat surprisingly, it is also the least trivial of the four. Here we give Alcuin's own solution, in the translation of John Hadley.

First of all, I and my sister would get into the boat and travel across; then I'd send my sister out of the boat and I would cross the river again. Then the sisters who had stayed on the bank would get into the boat. These having reached the other bank and disembarked, my sister would get into the boat and bring it back to us. She having got out of the boat, the other two men would board and go across. Then one of them with his sister would cross back to us in the boat. Then I and the man who had just crossed would go over again, leaving our sisters behind. Having reached the other side, one of the two women would take the boat across, and having picked up my sister could take the boat across to us. Then he, whose sister remained on the other side, would cross in the boat and bring her back with him. And that would complete the crossing without anything untoward happening. □

Notes. Once again, one can solve this problem in a totally mindless way, just listing the states we can get to if we are in a certain state, and then following a route from $[AaBbCc, \emptyset)$ to $(\emptyset, AaBbCc]$, with the brackets indicating the position of the boat. The solution above is the following sequence, where $\leftarrow \{Bb\} \leftarrow$ means that the boat takes B and his sister b from the right bank to the left bank:

$$[AaBbCc, \emptyset) \rightarrow \{Aa\} \rightarrow (BbCc, Aa] \leftarrow \{A\} \leftarrow [ABbCc, a) \rightarrow$$
$$\{bc\} \rightarrow (ABC, abc] \leftarrow \{a\} \leftarrow [AaBC, bc) \rightarrow \{BC\} \rightarrow (Aa, BbCc] \leftarrow$$
$$\{Bb\} \leftarrow [AaBb, Cc) \rightarrow \{AB\} \rightarrow (ab, ABCc] \leftarrow \{c\} \leftarrow [abc, ABC) \rightarrow$$
$$\{ac\} \rightarrow (b, AaBCc] \leftarrow \{B\} \leftarrow [Bb, AaCc) \rightarrow \{Bb\} \rightarrow (\emptyset, AaBbCc].$$

For more information about crossing problems, both mathematical and historical, see the lovely article Pressman and Singmaster published in 1989. In particular, as they remark, in 1879, a young student, Cadet de Fontenay, observed that four or more couples could cross if they could make use of an island in the river. For $n \geq 4$ couples, de Fontenay gave a solution with $8n - 6$ crossings. Pressman and Singmaster showed that this was best if a two-person boat is used and no bank-to-bank crossings are allowed. They also proved that if bank-to-bank crossings are allowed then for $n > 4$ the minimum is $4n + 1$. Pressman and Singmaster also showed that a solution of the original problem with eleven crossings, as above, needs three people who can row; if there are only two rowers, the minimum is 13 crossings.

In the 19th century several variants of these problems appeared, with a number of memorable names: jealous husbands, missionaries and cannibals, men travelling with their harems, masters and dishonest servants, servants and vicious masters, etc. Here is a basic variant: n missionaries and n cannibals must cross a river with an island, using a two-person boat, in such a way that the cannibals never outnumber the missionaries. Pressman and Singmaster proved that if bank-to-bank crossings are forbidden then the minimum is $8n - 6$, and if $n \geq 3$ and bank-to-bank crossings are allowed then the minimum is $4n - 1$. The reader may find it amusing to prove this for $n = 3$.

References

Alcuin of York, Propositiones ad acuendos juvenes (Problems to sharpen the young). An annotated translation of the oldest mathematical problem collection in Latin, translated by J. Hadley, commentary by D. Singmaster and J. Hadley, *Math. Gaz.* **76** (1992) 102–126.

Fraley, R., K.L. Cooke and P. Detrick, Graphical solution of difficult crossing puzzles, *Math. Mag.* **39** (1966) 151–157.

Pressman, I. and D. Singmaster, The jealous husbands and the missionaries and cannibals, *Math. Gaz.* **73** (1989) 73–81.

Schwartz, B.L., An analytic method for the "difficult crossing" puzzles, *Math. Mag.* **34** (1960/61) 187–193.

116. Fibonacci and a Medieval Mathematics Tournament

Two of the many questions in the mathematics tournaments to test the skill of Leonardo Fibonacci in the presence of Emperor Frederick II in Pisa in 1225 were as follows.

(i) Find a rational number of which the square, when either increased or decreased by 5, would remain a square.

(ii) Three men, A, B, C, possess a sum of money u, their shares being in the ratio 3 : 2 : 1. A takes away x, keeps half of it, and deposits the remainder with D; B takes away y, keeps two-thirds of it, and deposits the remainder with D; C takes away all that is left, z, keeps five-sixths of it, and deposits the remainder with D. This deposit with D is found to belong to A, B and C in equal proportions. Find u, x, y and z.

Solution. (i) Fibonacci's answer was 41/12. Then $(41/12)^2 + 5 = (1681 + 720)/12^2 = (49/12)^2$ and $(41/12)^2 - 5 = (1681 - 720)/12^2 = (31/12)^2$.

(ii) The pot of money is $u = x + y + z$, of which A, B and C own $u/2$, $u/3$ and $u/6$. After the division procedure above, A, B and C have $x/2$, $2y/3$ and $5z/6$, so D holds $u/2 - x/2$ of A's money, $u/3 - 2y/3$ of B's and $u/6 - 5z/6$ of C's. Consequently,

$$\frac{x+y+z}{2} - \frac{x}{2} = \frac{x+y+z}{3} - \frac{2y}{3} = \frac{x+y+z}{6} - \frac{5z}{6}.$$

Fixing z, and solving the two equations above for x and y, we find that $x = 33z$ and $y = 13z$. In particular, $u = 47$, $x = 33$, $y = 13$ and $z = 1$ is a solution. □

Notes. For an account of this tournament, see pp. 169–170 of the book of Walter William Rouse Ball (1850–1925), usually known as W.W. Rouse Ball. A mathematician, lawyer and enthusiastic amateur magician, he was Second Wrangler in 1874, and then First Smith Prizeman, and from 1875 till his death he was a Fellow of Trinity College, Cambridge. He wrote a fascinating book on 'Mathematical Recreations', and another on the history of mathematics. As a long-time tutor (in the days when tutors were much more important and had much more power than today), he looked after his students extremely well: he even had a little building erected in his garden to house a billiards table for them.

He was a generous benefactor of Trinity College and the universities of Oxford and Cambridge: he founded a Rouse Ball Professorship of Mathematics in both places, and a Rouse Ball Professorship of English Law at Oxford. The list of people who occupied these chairs could hardly be more distinguished, e.g. in Cambridge J.E. Littlewood was the first Rouse Ball Professor of Mathematics; he was followed by A.S. Besicovitch, Harold Davenport, John G. Thompson, Nigel Hitchin and Sir Timothy Gowers.

Reference

Ball, W.W. Rouse, *A Short Account of the History of Mathematics*, Macmillan (1888).

117. Triangles and Quadrilaterals – Regiomontanus

(i) *Let ABC be a triangle such that the foot D of its altitude from A is on the side BC. Suppose that AC − AB = 3, DC − DB = 12 and AD = 10. Then the base BC has length $\frac{\sqrt{321}}{3}$.*

(ii) *Suppose that there are quadrilaterals with side lengths a, b, c and d in a cyclic order. Then one can construct the one which is inscribed in a circle.*

Solution. These problems, needing only simple-minded high-school mathematics, are bound to be beneath the dignity of the reader, but we shall still spell out two pedestrian, head-on solutions before we say a few words about the history of the problems.

(i) Set $x = BD$ and $y = AB$, so that $AC = y + 3$ and $DC = x + 12$, as in Figure 54. Since ADB and ADC are right-angled triangles (with the right

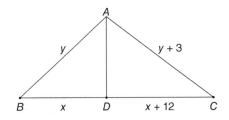

Figure 54 The notation used.

angles at D), by Pythagoras's theorem we have

$$x^2 + 10^2 = y^2 \qquad \text{and} \qquad (x + 12)^2 + 10^2 = (y + 3)^2.$$

On taking the difference of the two equations, we find that

$$y = 4x + \frac{45}{2}.$$

Substituting this into the first equation, we find that

$$x^2 + 100 = 16x^2 + 180x + \frac{2025}{4}.$$

Solving this quadratic equation, we get $x = \frac{\sqrt{321}}{6} - 6$, which tells us that $BC = \frac{\sqrt{321}}{3}$.

(ii) If a polygon with side lengths $a_1, \ldots, a_k$ has a circumcircle of radius r then for every order of these lengths there is a unique cyclic polygon with these sides (and the circumcircle has the same radius r). In particular, given lengths $a_1, \ldots, a)k$ with max $a_i = a_1$, there is a cyclic polygon with these side lengths if and only if $a_1 < a_2 + \cdots + a_k$; also, this cyclic polygon is unique (i.e. the radius of its circumcircle is unique).

This tells us that we may assume that $a \geq b \geq c \geq d$ and $b + c + d > a$. Let $ABCD$ be the unique cyclic quadrilateral with $AB = a$, $BC = b$, $CD = c$ and $DA = d$. Let $\beta \leq \pi/2$ be the angle at B; then the angle at D is $\pi - \beta$. We shall calculate the length of the diagonal from two sides, from ABC and ADC, and equate the two expressions to determine β.

From the triangle ABC, with CF the height from C, as in Figure 55, we see

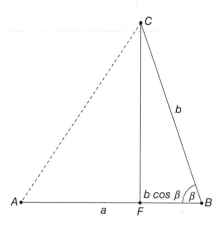

Figure 55 The notation in the triangle ABC: F is the foot of the height from C.

that

$$AC^2 = AF^2 + FC^2 = (a - b\cos\beta)^2 + b^2 \sin^2\beta = a^2 + b^2 - 2ab\cos\beta;$$

analogously, from the triangle CFA we get

$$AC^2 = c^2 + d^2 - 2cd\cos(\pi - \beta) = c^2 + d^2 + 2cd\cos\beta.$$

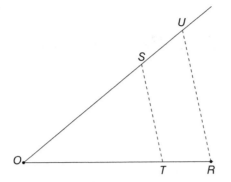

Figure 56 $OR = r, OT = t$ and $OU = u$, and S is on the half-line OU, with ST parallel to RU.

Consequently,

$$\cos \beta = \frac{a^2 + b^2 - c^2 - d^2}{2(ab + cd)}. \tag{1}$$

All that remains is to construct the angle β from (1), in which $\cos \beta$ is given as the ratio of two combinations of areas of squares and rectangles. First, change all areas to areas of rectangles with one side a, or whatever we like, so that the ratio in (1) appears as the ratio of two lengths. As a general step, given r, t and u, set $OR = r$, $OT = t$ and $OU = u$, as in Figure 56. Construct S with ST parallel to RU and set $s = ST$. Then $rs = tu$: a rectangle with sides t and u has been turned into a rectangle of the same area with one side r. Second, having turned $\cos \beta$ into the ratio of two lengths, just recall the definition of cosine. □

Notes. The problems are from *De Triangulis*, the first systematic exposition of trigonometry, written by Johann Regiomontanus in 1464 (and printed in Nürnberg in 1533). Regiomontanus was born in Königsberg in 1436 as Johannes Müller but, as was customary in those days, wrote under a Latin name (which he took from his birthplace).

Regiomontanus's proofs were somewhat convoluted, which is not very surprising since he had to express everything in words. The formulae we have used were not available in 1464. Needless to say, in the first part Regiomontanus used the numbers 3, 12 and 10 only to illustrate his method: as we have seen, these numbers did not even lead to a particularly nice solution. What is remarkable in the deductions above is that Regiomontanus did not hesitate to use algebraic methods to solve a geometric problem.

118. The Cross-Ratios of Points and Lines

(i) *Let A, B, C and D be four points on a line ℓ, and let O be a fifth point not on ℓ. Then*

$$O[ABCD] = [A, B; C, D].\tag{1}$$

(ii) *Let A, B, C, D, O and O' be six points on a circle. Then*

$$O[ABCD] = O'[ABCD].\tag{2}$$

(iii) *Let* $[A, B; C, D] = [A', B; C, D]$, *where the two sets of arguments are collinear, and A, C, D and A', C, D are in the same order. Then* $A = A'$. *Furthermore, the analogous assertion holds for lines and their cross-ratios as well.*

Proof. (i) Write a, b, c and d for the lines determined by the segments OA, OB, OC and OD, as in Figure 57, and write (a), (b), (c) and (d) for the angles between $ℓ$ and the lines a, b, c and d. The only condition is that these angles are taken to be between 0 and π; we may take (a) or $\pi - (a)$, etc. Also, let (ac) be the angle of the triangle OAC at O, (ad) the angle of the triangle OAD at O, etc. Then, by the Sine Rule,

$$AC = \sin(ac)\,\frac{OA}{\sin(c)}, \qquad\qquad BC = \sin(bc)\,\frac{OB}{\sin(c)},$$

$$AD = \sin(ad)\,\frac{OA}{\sin(d)}, \qquad\qquad BD = \sin(bd)\,\frac{OB}{\sin(d)}.$$

Consequently,

$$[A, B; C, D] = \frac{AC}{AD}\left/\frac{BC}{BD}\right. = \frac{\sin(ac)}{\sin(ad)}\left/\frac{\sin(bc)}{\sin(bd)}\right. = [a, b; c, d] = O[ABCD],$$

proving (1).

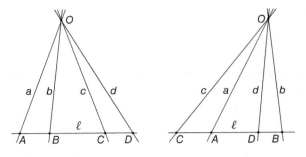

Figure 57 Two arrangements of the points A, B, C, D and lines a, b, c, d.

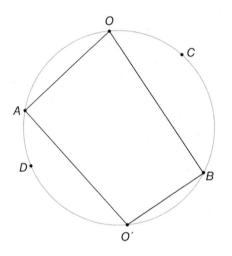

Figure 58 Our six points on a circle.

(ii) Two angles of the type $\angle OB$ and $\angle O'B$ are either equal or sum to π, so the sine function has the same value on them, implying (2).

(iii) All we need is that, for C and D fixed and $A \neq C, D$, the ratio AC/AD takes every value at most once. In fact, as $A \neq C, D$ and none of our points is at infinity, AC/AD takes every value other than $0, 1$ and ∞ exactly once. Indeed, assuming, as we may, that ℓ is the x-axis, $C = (0,0)$, $D = (1,0)$ and $A = (x,0)$, $x \neq 0, 1$, we have $AC/AD = x/(x-1) \neq 0, 1$. Now, if $x/(x-1) = t \neq 0, 1$ then $x = t/(t-1) \neq 0, 1, \infty$. $\qquad\square$

Notes. The way we have introduced the cross-ratio was rather pedestrian: let us repeat it here in a slightly more sophisticated way, identifying the real plane with $\mathbb{C}$, the field of complex numbers. Given four complex numbers, $z_1, \ldots, z_4 \in \mathbb{C}$, their *cross-ratio* is

$$[z_1, z_2; z_3, z_4] = \frac{(z_1 - z_3)(z_2 - z_4)}{(z_1 - z_4)(z_2 - z_3)}.$$

Clearly, the cross-ratio of complex numbers (points in the complex plane) depends on their order in the expression above, so *a priori* there are twenty-four different values for a set of four numbers. However, four permutations (including the identity) leave the fraction unchanged:

$$[z_1, z_2; z_3, z_4] = [z_2, z_1; z_4, z_3] = [z_3, z_4; z_1, z_2] = [z_4, z_3; z_2, z_1],$$

so there can be at most six different values. In fact, if $[z_1, z_2; z_3, z_4] = \lambda$, then the twenty-four permutations produce λ, $1/\lambda$, $1 - \lambda$, $1/(1 - \lambda)$, $1 - 1/\lambda = (\lambda - 1)/\lambda$, and $\lambda/(\lambda - 1)$. Needless to say, there need not be six different values: e.g. when $\lambda = e^{\pm i\pi/3}$ then there are only two values, $e^{i\pi/3}$ and $e^{-i\pi/3}$. From here, we define the cross-ratio of lines as before.

Strictly speaking, the cross-ratio is defined on the Riemann sphere $\mathbb{C} \cup \infty$. For example, $[z_1, z_2; z_3, \infty] = (z_1 - z_3)/(z_2 - z_3)$ and so $[z, 1; 0, \infty] = z$. It is easily seen that the cross-ratio is preserved by every Möbius (or fractional linear) transformation, i.e. transformation of the form $z \mapsto (az + b)/(cz + d)$, with $ad - bc \neq 0$.

Given three different complex numbers, $z_2, z_3, z_4 \in \mathbb{C}$, the cross-ratio $g(z) = [z, z_2; z_3, z_4]$ is precisely a Möbius transformation:

$$g(z) = ((z_2 - z_4)z + (z_3 z_4 - z_2 z_3)) / ((z_2 - z_3)z + (z_3 z_4 - z_2 z_4)).$$

To check that g is a Möbius transformation, note that

$$(z_2 - z_4)(z_3 z_4 - z_2 z_4) = (z_2 - z_3)(z_3 z_4 - z_2 z_3)$$

holds if and only if

$$(z_2 - z_3)(z_3 - z_4)(z_4 - z_2) = 0.$$

Since the inverse of a Möbius transformation $(az + b)/(cz + d)$ is the Möbius transformation $(-dz + b)/(cz - a)$, (iii) follows.

The cross-ratio of four distinct points (i.e. complex numbers) is real if and only if the points are on a conic section (ellipse, hyperbola or parabola). Indeed, two conics can be mapped into each other by a Möbius transformation, and the cross-ratio of four points, three of which are collinear, is real if and only if all four points are collinear. Hence (ii) holds not only for circles but for conics as well.

The cross-ratio was first defined early in the 19th century by the French Lazare Nicolas Marguérite Carnot (1753–1823), Charles Julien Brianchon

(1783–1864) and the German August Ferdinand Möbius (1790–1868). It gained prominence only later, in the work of the brilliant Swiss geometer Jakob Steiner (1796–1863) and the outstanding British algebraists, Arthur Cayley (1821–1896) and William Kingdon Clifford (1845–1879).

Cayley, an extraordinarily prolific mathematician and the greatest pure mathematician in Britain in the 19th century, was a student of Trinity College, Cambridge: he graduated as Senior Wrangler in 1842 and went on to a Fellowship of Trinity.

Although for fourteen years he worked as a lawyer, even during that time he wrote about 250 mathematical papers. In 1863 he returned to Cambridge as the Sadleirian Professor and a few years later, till the end of his life, he was again a Fellow of Trinity College.

Clifford was also a student of Trinity: he graduated as Second Wrangler in 1867 and went on to a Fellowship in the college. From 1871 till his death seven and a half years later, he was a professor at University College London. It was Clifford who invented the name 'cross-ratio'.

For an excellent introduction to projective geometry (and so the use of cross-ratios) see Baltus's 2020 book on conic sections.

References

Baltus, C., *Collineations and Conic Sections – An Introduction to Projective Geometry in its History*, Springer (2020).

Brianchon, C.J., *Mémoire sur les Lignes du Second Ordre*, Bachelier (1817).

Möbius, A., *Der Barycentrische Calcul: ein neues Hülfsmittel zur analytischen Behandlung der Geometrie*, Barth (1827).

119. Hexagons in Circles: Pascal's Hexagon Theorem – Take One

Let A, B, C, D, E and F be points on a circle and suppose that the opposite sides of the 'hexagon' ABCDEF meet in G, H and I. Then the points G, H and I are on a line.

Proof. We shall make use of the results concerning the cross-ratio in the previous problem. Recall that, given four points, P, Q, R, S and O, such that no triple containing O is collinear, $O[PQRS]$ denotes the cross-ratio of the four lines through OP, OQ, OR and OS.

The notation we shall use is given in Figure 59: we have marked two additional points: J, the intersection of the sides CD and EF, and K, the intersection of the sides DE and FA.

We claim that

$$I[EJFH] = C[EJFH] = C[EDFB] = A[EDFB]$$
$$= A[EDKG] = I[EDKG] = I[EJFG].$$

To prove this claim, we prove the six equalities above, but not in the order we have them there. First note that $I[EJFH] = C[EJFH]$ and $A[EDKG] = I[EDKG]$, since E, J, F and H are collinear, and so are E, D, K and G. Second, $C[EJFH] = C[EDFB]$ since they denote the cross-ratio of the same set of lines; also, $A[EDFB] = A[EDKG]$ and $I[EDKG] = I[EJFG]$ for the same reason. Third, $C[EDFB] = A[EDFB]$ since the six points A, B, C, D, E and F are on a conic. Putting these equalities together, the claim follows.

The equality $I[EJFH] = I[EJFG]$ we have just proved implies Pascal's theorem. Indeed, as the cross-ratio of the lines IE, IJ, IF and IH is the same as the cross-ratio of IE, IJ, IF and IG, the lines IG and IH are identical, i.e. I, G and H are collinear. □

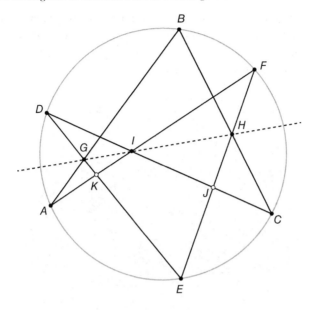

Figure 59 Our notation.

Notes. Blaise Pascal (1623–1662), one of the greatest mathematicians ever, proved his *Hexagon Theorem* above in 1640, when he was sixteen. The proof above has nothing to do with Pascal, as cross-ratio had not yet been introduced, but this is the proof I learned as a student and find simplest in the sense that it needs the least amount of ingenuity. The use of the cross-ratio also tells us that we have proved more than stated: the assertion holds if instead of a circle we take a conic section (ellipse, parabola or hyperbola). Often this more general theorem is called Pascal's Hexagon Theorem. Not surprisingly, this theorem is in fact also a simple exercise in elementary algebraic geometry.

Pascal's Hexagon Theorem is a stepping stone to *Pascal's Hexagrammum Mysticum*. Six points on a conic section can be connected into a 'hexagon' in $5!/2 = 60$ different ways. Applying Pascal's Hexagon Theorem to these hexagons, we get 60 lines, each going through three points: this configuration is Pascal's Hexagrammum Mysticum, which in the 19th century was a celebrated result.

Pascal was not only a mathematical prodigy, who worked mostly in geometry, but also a physicist, a philosopher, a technologist who designed and constructed a calculating machine, and planned and inaugurated the first public service in Paris, a probabilist before probability theory got off the ground, a scientist influencing economics and social sciences, and a major Catholic theologian.

Sadly, he never published his life-work on Catholic philosophy, but left behind fragments that became known as his *Pensées*: a collection of loosely connected thoughts, especially about faith in God; the language in this work is often considered to be French at its very best.

There was another exceptional – but today largely forgotten – Pascal, the Hungarian film director Gabriel Pascal (1894–1954). Needless to say, Pascal was not his family name, which he *never* revealed: his origin was shrouded in mystery which, according to his wife, he enjoyed thickening with contradictory remarks. [I am unwilling to believe the information on Wikipedia.] His life, especially in his youth, was extraordinarily romantic.

Gabriel Pascal was the only film director with whom G.B. Shaw (1856–1950) collaborated. Shaw thought very highly of Pascal: 'Gabriel Pascal is one of those extraordinary men who turn up occasionally – say once in a century – and may be called godsends in the arts to which they are devoted.' Their *Caesar and Cleopatra*, shot in 1945, is a sheer delight.

Reference

Pascal, B., *Pensées* (1670). Translated with an Introduction by A.J. Krailsheimer, Penguin Classics (1995).

120. Hexagons in Circles: Pascal's Theorem – Take Two

Let the vertices of a hexagon lie on a circle, with the three pairs of opposite sides intersecting. Then the points of intersection are collinear.

Proof. The detailed hint we have given goes a long way towards a proof, but here we shall give all the details.

As we shall have points of three different types, we choose our notation accordingly. Let $A_0 A_1 \cdots A_5$ be our 'hexagon' inscribed in a circle, the 'first circle'. Let $A_0 A_1$ intersect $A_3 A_4$ in the point P_0, $A_1 A_2$ intersect $A_4 A_5$ in P_1, and $A_2 A_3$ intersect $A_5 A_0$ in P_2. Consider a 'second circle' through A_1, A_4 and P_1. Let this circle intersect the line through A_4, A_3 and P_0 in B_3 in addition to A_4 and let it also intersect the line through A_0, A_1 and P_0 in B_0 in addition to A_1. Similarly, let this circle intersect the line through A_3, A_4 and P_0 in B_3 in addition to A_4. See Figures 60 and 61 for two arrangements.

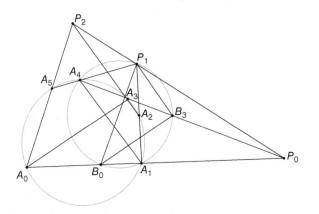

Figure 60 A hexagon inscribed in a circle with its additional points.

315

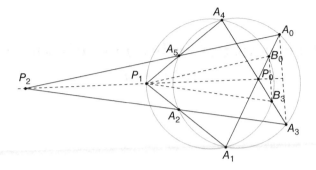

Figure 61 The triangles $A_0A_3P_2$ and $B_0B_3P_1$ are perspective from P_0.

To prove Pascal's theorem, it suffices to show that the triangles $A_0A_3P_2$ and $B_0B_3P_1$ are homothetic. Indeed, since the lines A_0B_0 and A_3B_3 meet in P_0, these triangles are perspective from P_0: in particular, the points P_0, P_1 and P_2 are collinear, as claimed.

That the two required triangles are homothetic follows from chasing angles, using the fact that, in a circle, two angles on the same arc are equal and, on complementary arcs, sum to 2π. Accordingly, different arrangements need slightly different arguments – here we shall consider the arrangement in Figure 61. First, taking the arc A_1B_3 of the second circle, and then the arc A_1A_3 of the first circle, we have

$$\angle_1 B_0B_3 = \angle_1 A_4B_3 = \angle_1 A_4A_3 = \angle_1 A_0A_3,$$

so the sides A_0A_3 and B_0B_3 are parallel.

Second, considering the arc A_4P_1 of the second circle, and then the arc A_2A_4 of the first circle, we find that

$$\angle_4 B_3P_1 = \angle_4 A_1P_1 = \angle_4 A_1A_2 = \angle_4 A_3A_2 = \angle_4 A_3P_2.$$

As the points A_4, A_3 and B_3 are collinear, the sides P_1B_3 and P_2A_3 are parallel.

Finally, considering the arc P_1B_3 of the second circle, and the arc A_3A_5 of the first circle, we see that

$$\angle_1 B_0B_3 = \angle_1 A_4B_3 = \angle_5 A_4A_3 = \angle_5 A_0A_3 = \angle_2 A_0A_3.$$

Since the sides A_0A_3 and B_0B_3 are known to be parallel, so are the sides A_0P_2 and B_0P_1. Hence, the triangles $A_0A_3P_2$ and $B_0B_3P_1$ are indeed homothetic, completing our proof.   □

Notes. This very simple proof of Pascal's theorem was found by van Yzeren in 1993. It's not that it was easy to find this proof – far from it! I consider the use of the 'second circle' a stroke of genius.

It is not impossible that this was the proof Pascal gave of his theorem: as van Yzersen wrote, 'Whether Pascal gave his proof is open to debate, but it seems that this proof has not turned up for 350 years.' Concerning this proof, H.S.M. "Donald" Coxeter (1907–2003) wrote this to van Yzeren: 'It is indeed remarkable that this elegant proof was not found in 350 years, and also somewhat remarkable that Guggenheimer came close to it in 1967 and then felt obliged to introduce a peculiar lemma.'

Although in 1925 Coxeter was offered a place at King's College, Cambridge to read for the Mathematical Tripos, he decided to wait a year in order to get into Trinity College, Cambridge with a scholarship. (Mathematicians are not competitive!) In 1928 he came top of the Tripos, so would have been the Senior Wrangler if Hardy and his allies had not abolished the order of merit. Three years later he was awarded his PhD and a Title A (Junior Research) Fellowship in Trinity College. After two long visits to Princeton, he moved to the University of Toronto, where he stayed till the end of his long life. He was often called "the man who saved geometry" and "the twentieth century's greatest geometer"; his *Coxeter groups* have found wide applications.

References

Guggenheimer, H.W., *Plane Geometry and its Groups*, Holden-Day, Inc (1967).

Roberts, S. and A.I. Weiss, Obituary: Harold Scott Macdonald Coxeter, FRS, 1907–2003, *Bull. Lond. Math. Soc.* **41** (2009) 943–960.

van Yzeren, J., A simple proof of Pascal's hexagon theorem, *Amer. Math. Monthly* **100** (1993) 930–931.

121. A Sequence in $\mathbb{Z}_p$

For a prime p, let $a_1, \ldots, a_{p-1}$ be a sequence of elements of $\mathbb{Z}_p$ such that $\sum_{i \in I} a_i \neq 0$ whenever I is a non-empty set of indices. Then this sequence is constant: $a_1 = \cdots = a_{p-1}$.

Proof. For $j = 1, \ldots, p-1$, set $s_j = \sum_{i=1}^{j} a_i$. If we had $s_j = 0$ for some j, then $I = \{1, \ldots, j\}$ would do, and if we had $s_j = s_k$ for some $1 \leq j < k \leq p-1$, then $I = \{j+1, \ldots, k\}$ would do. Hence, neither of these events happens, and so the sequence $s_1 = a_1, s_2 = a_1 + a_2, \ldots, s_{p-1} = a_1 + \cdots + a_{p-1}$ is a permutation of the non-zero elements $1, 2, \ldots, p-1$ of $\mathbb{Z}_{p-1}$. The same holds for the sequence $s'_1 = a_2, s'_2 = a_1 + a_2 = s_2, s'_3 = s_3, \ldots, s'_{p-1} = s_{p-1}$ obtained from the sequence $a_2, a_1, a_3, \ldots, a_{p-1}$ arising from the original sequence by interchanging the first two terms. Since the last $p-2$ elements of the permutations (s'_j) and (s_j) of $\{1, 2, \ldots, p-1\}$ are identical, the first elements are also equal: $s'_1 = a_2 = s_1 = a_1$, telling us that $a_1 = a_2$. Since this holds for any rearrangement of the sequence (a_i), all the a_i are the same. $\qquad \square$

Notes. The assertion is trivially false for shorter sequences, as shown by the sequence $1, 1, \ldots, 1, 2$ of length $p-2$. (Here $p-3$ of the terms are equal to 1, and one is 2.)

122. Elements of Prime Order

Let G be a finite group whose order N is divisible with a prime p. Then the number of elements of order p in G is congruent to −1 modulo p.

Proof. Let K be the subset of $G^p = G \times \cdots \times G$ consisting of the vectors $\mathbf{v} = (x_1, \ldots, x_p)$ such that $x_1 \cdots x_p = e$, where e is the identity of G. Clearly, $|K| = N^{p-1}$ since the first $p - 1$ coordinates of $\mathbf{v} = (x_1, \ldots x_p) \in K$ can be chosen in any way we like, while x_p is determined by these elements: $x_p = (x_1 \cdots x_{p-1})^{-1}$. In particular, $|K|$ is a multiple of p.

Note that if $x_1 \cdots x_p = e$, i.e. $x_2 \cdots x_p = x_1^{-1}$, then $x_2 \cdots x_p x_1 = x_1^{-1} x_1 = e$. In other words, if $(x_1, \ldots, x_p)$ is in K then so is its cyclic permutation $(x_2, \ldots, x_p, x_1)$. This shows that the cyclic group $\mathbb{Z}_p$ acts on K by permutation, with $i \in \{0, 1, \ldots, p-1\}$ mapping $(x_1, \ldots, x_p)$ into $(x_{i+1}, x_{i+2}, \ldots, x_p, x_1, \ldots, x_i)$. Trivially, the set K is the disjoint union of orbits of various elements: the orbits of two elements are either identical or disjoint.

How large is the orbit $O_\mathbf{v}$ of a vector $\mathbf{v} = (x_1, \ldots, x_p)$? If not all the x_i are the same, then $|O_\mathbf{v}| = p$: no non-trivial cyclic permutation leaves $\mathbf{v}$ invariant. Furthermore, a constant vector $\mathbf{v}_x = (x, \ldots, x)$ belongs to K if and only if $x^p = e$, and in this case the orbit of $\mathbf{v}_x$ is trivial: $O_{\mathbf{v}_x} = \{\mathbf{v}_x\}$. In particular, the number of trivial (singleton) orbits, t, say, is one more than the number of elements of order p. Since $|K|$ is a multiple of p and every non-trivial orbit has p elements, t is indeed divisible by p, completing the proof. □

123. Flat Triangulations

Call a triangulation of a polygon flat if every internal vertex has degree 6, as in Figure 62, and write $f(n)$ for the maximal number of triangles in a flat triangulation of an n-gon. Then $f(n) \leq n^2/6$, with equality whenever n is a multiple of 6.

Proof. We shall prove the assertion by induction on n. One can show that for $3 \leq n \leq 5$ we have $f(n) = n - 2$, so our inequality does hold. Suppose then that $n \geq 6$, $f(n) \geq 7$ and $f(m) \leq m^2/6$ for $m \leq n - 1$.

Let T_n be a flat triangulation of an n-gon P_n with m internal vertices and $k = f(n)$ triangles, and let $d_1, \dots, d_n$ be the degrees of the vertices of P_n. We may assume that every d_i is at least 3, since otherwise the deletion of a suitable triangle of T_n leaves us with a flat triangulation of an $(n-1)$-gon, so by the induction hypothesis, $f(n) \leq 1 + f(n-1) \leq 1 + (n-1)^2/6 < n^2/6$.

We claim that

$$\sum_{i=1}^{n} d_i = 4n - 6. \tag{1}$$

To see (1), extend T_n to a triangulation T'_n of the sphere by the addition of a vertex outside the polygon P_n, and joining it to the vertices of P_n. Write V for the number of vertices of T'_n, E for the number of its edges and F for the number of faces of its map, so that $V = n + m + 1$ and $2E = 6m + \sum_{i=1}^{n} d_i + 2n$. By Euler's Polyhedron Formula we have $V + F = E + 2$; also, as every face of T'_n is a triangle, $2E = 3F$. Hence, $V + 2E/3 = E + 2$, so $2E = 6V - 12$, i.e. the sum of the degrees of the vertices of T'_n is

$$6m + \sum_{i=1}^{n} d_i + 2n = 6(n + m + 1) - 12.$$

Rearranging this, we obtain (1).

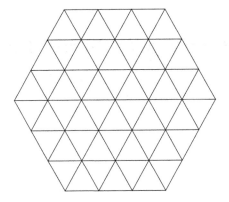

Figure 62 A triangulation of a (degenerate) 18-gon with 54 triangles.

Now, to make use of the induction hypothesis, peel off T_n all the triangles that meet at least one vertex of P_n. The remaining triangles of T_n form flat triangulations of some polygons, with $n_1, \ldots, n_r$ vertices, say. For every edge xy of these polygons, there is a triangle xyz of the triangulation T_n such that z is a vertex of P_n but neither xz nor yz is a side of P_n. Since a vertex z of P_n of degree $d(z) \geq 3$ is in $d(z) - 3$ such triangles,

$$\sum_{i=1}^{r} n_i \leq \sum_{i=1}^{n} (d_i - 3) = n - 6.$$

In view of this, the number of triangles we have peeled off T_n is at most

$$n - 6 + n = 2n - 6,$$

as T_n has n triangles on the sides of P_n. Consequently, by the induction hypothesis,

$$k = f(n) \leq 2n - 6 + f(n - 6) \leq 2n - 6 + (n - 6)^2/6 \leq n^2/6,$$

as required.

Finally, if $n = 6\ell$ then a hexagonal part of the triangular lattice (with ℓ triangles on each side of the hexagon) is a flat triangulation of an n-gon with $n^2/6$ triangles, as in Figure 62.   □

124. Triangular Billiard Tables

Let ABC be a triangular billiard table with acute angles, and let P, Q and R be points on the sides BC, CA and AB. Then a billiard ball launched from P towards Q describes the 3-periodic polygonal path PQRPQR··· if and only if P, Q and R are the feet of the heights of the triangle ABC.

Proof. Let us write α, β and γ for the angles of the triangle ABC.

(i) Suppose our billiard ball runs along the triangular path $PQRPQR\cdots$ ad infinitum. Write φ, ψ and ξ for the angles of reflection, so that $\angle PR = \angle PQ = \varphi$, etc., as in Figure 63. Then the triangles ARQ, BPR and CQR tell us that

$$\alpha + \xi + \psi = \beta + \varphi + \xi = \gamma + \psi + \varphi = \pi,$$

so $\varphi = \alpha$, $\psi = \beta$ and $\xi = \gamma$.

In the rectangle $ABPQ$ the angle at A is α, and the angle at P is $\pi - \varphi = \pi - \alpha$. Consequently, $ABPQ$ is a cyclic quadrilateral (i.e. a quadrilateral written in a

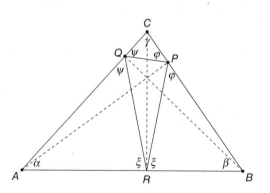

Figure 63 The distribution of the angles in the triangle ABC when the billiard ball runs along the path $PQRPQR\ldots$.

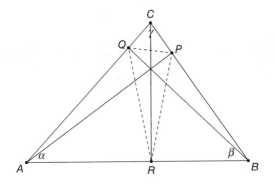

Figure 64 The points P, Q and R are the feet of the heights.

circle). Therefore $\angle PB = \angle QA$. Analogously, $\angle QC = \angle RB$ and $\angle RA = \angle PC$. Of these six angles, the two at P sum to π, the two at Q sum to π, and the two at R sum to π. Hence, we have

$$\angle PB = \angle QA = \pi - \angle QC = \pi - \angle RB$$
$$= \angle RA = \angle PC = \pi - \angle PB,$$

so all six angles are $\pi/2$. Thus P, Q and R are indeed the feet of the heights of our triangle ABC. (ii) Conversely, let P, Q and R be the feet of the heights of ABC from A, B and C, as in Figure 64. Then the quadrilaterals $ABPQ$, $BCQR$ and $CARP$ are cyclic since, for example, $\angle PB = \angle QB = \pi/2$. (Needless to say, it does not matter that these two angles are $\pi/2$, only that they are equal.) Now, as $ABPQ$ is cyclic, $\angle QC = \angle BA = \beta$, and as $BCQR$ is cyclic, $\angle QA = \angle BR = \beta$. Thus $\angle QC = \angle QA$, so the billiard ball launched from P towards Q bounces to R. Analogously, from R it bounces to P, then to Q, etc. Thus the ball runs along the path $PQRPQR\ldots$, as claimed. $\qquad\square$

125. Chords of an Ellipse: The Butterfly Theorem

Let AB be a chord of an ellipse with midpoint M, and let PQ and RS be two other chords through M. Denote by T and U the intersections of the chords PS and RQ with AB. Then M is the midpoint of the segment TU.

Proof. The essential part of this theorem is a baby version of the proof of Pascal's theorem in Problem 119. As there, we use the abbreviated notation $[ABCD]$ for the cross-ratio of the points A, B, C and D. It does not help us at all, but note that we may assume that our ellipse is in fact a circle, as in Figure 65.

Here is then the proof. Note first that

$$[A, T; M, B] = S[ATMB] = S[APRB] = Q[APRB] = Q[AMUB]$$
$$= [A, M; U, B] = [B, U; M, A].$$

Indeed, the first equality is just the connection between the cross-ratios of points and lines, as is the penultimate equality; the second and fourth hold since the

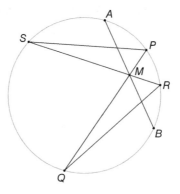

Figure 65   A butterfly in a circle.

lines in question are identical; and the third holds since not only A, P, R and B are on the ellipse (or circle), but so are S and Q. The last equality is immediate from the definition of the cross-ratio:

$$\frac{(z_1 - z_3)(z_2 - z_4)}{(z_1 - z_4)(z_2 - z_3)} = \frac{(z_4 - z_2)(z_3 - z_1)}{(z_4 - z_1)(z_3 - z_2)}.$$

Finally, since $[z_1, u; z_3, z_4] = [z_1, w; z_3, z_4]$ implies that $u = w$, and AMB is congruent to BMA, we find that $ATMB$ is congruent to $BUMA$. Hence M is indeed the midpoint of TU.   □

Notes. This problem has gained notoriety because it was given as an entrance examination question to 'special students', most of whom were Jewish, at Moscow State University during the communist era so as to deny them places at the university.

References

Shen, A., Entrance examinations to Mekh-Mat, *Math. Intell.* **16** (1994) 6–10.

Vardi, I., Mekh-Mat entrance examination problems. In *You Failed Your Math Test, Comrade Einstein*, M. Shifman (ed.), World Scientific (2005).

Vershik, A., Admissions to the mathematics faculty in Russia in the 1970s and 1980s, *Math. Intell.* **16** (1994) 4–5.

126. Recurrence Relations for the Partition Function

The partition function satisfies the following recurrence relations:

$$np(n) = \sum_{k=1}\sum_{v=1} vp(n - kv) \tag{1}$$

and

$$p(n) = p(n - 1) + p(n - 2) - p(n - 5) - p(n - 7) + p(n - 12) \pm \ldots$$
$$= \sum_{k=1}(-1)^{k+1}\left(p\left(n - \frac{k(3k - 1)}{2}\right) + p\left(n - \frac{k(3k + 1)}{2}\right)\right). \tag{2}$$

In the summations above, k takes all the positive values that do not give negative arguments for the partition function; thus, in the second summation k takes about $\sqrt{2n/3}$ values.

Proof. (i) To prove (1), consider all $p(n)$ identities given by the partitions. Thus, for $n = 5$, say, take

$$5 = 5,$$
$$5 = 4 + 1,$$
$$5 = 3 + 2,$$
$$5 = 3 + 1 + 1,$$
$$5 = 2 + 2 + 1,$$
$$5 = 2 + 1 + 1 + 1,$$
$$5 = 1 + 1 + 1 + 1 + 1.$$

Adding these $p(n)$ identities, the left-hand side is $np(n)$. But what is R, the value of the right-hand side?

Assuming that $1 \leq kv \leq n$, the number of partitions of n that contain a part v with multiplicity at least k is $p(n - kv)$. Also, if a partition λ of n contains

v with multiplicity exactly m, so that it contributes mv to R, then λ is counted among the partitions with multiplicities at least 1, at least 2, ..., at least m, so it contributes mv to $\sum_{k=1} \sum_{v=1} vp(n - kv)$. Consequently,

$$R = \sum_{k=1} \sum_{v=1} vp(n - kv),$$

completing our proof of (1).

Returning to our illustration with $n = 5$, there are three partitions with 2 and only one that contains 2 with multiplicity two: this gives $3 \cdot 2 + 1 \cdot 2 = 8$ as the contribution of 2 to R. Also, five partitions contain 1, three contain 1 at least twice, two contain 1 at least three times, one at least four times, and one five times. Hence 1 contributes $5 \cdot 1 + 3 \cdot 1 + 2 \cdot 1 + 1 \cdot 1 + 1 \cdot 1 = 12$ to R.

(ii) We know from Euler's simple result in Problem 26 that

$$\sum_{n=0}^{\infty} p(n)x^n = \prod_{k=1}^{\infty} (1 + x^k + x^{2k} + x^{3k} + \dots)$$

$$= \prod_{k=1}^{\infty} \frac{1}{1 - x^k}.$$

By Euler's Pentagonal Theorem in Problem 74,

$$\prod_{k=1}^{\infty} (1 - x^k) = 1 + \sum_{k=1}^{\infty} (-1)^k \left(x^{k(3k-1)/2} + x^{k(3k+1)/2} \right),$$

so

$$\left(\sum_{n=0}^{\infty} p(n)x^n \right) \left(1 + \sum_{k=1}^{\infty} (-1)^k \left(x^{k(3k-1)/2} + x^{k(3k+1)/2} \right) \right) = 1.$$

For $n \geq 1$ the coefficient of x^n on the left-hand side is

$$p(n) + \sum_{k=1}^{\infty} (-1)^k \left(p\left(n - \frac{k(3k-1)}{2}\right) + p\left(n - \frac{k(3k+1)}{2}\right) \right),$$

completing the proof of (2). □

Notes. The second recurrence is equivalent to Euler's Pentagonal Theorem: this is the recurrence Major MacMahon used to calculate by hand many values of the partition function. The first recurrence gives a much slower method, but it does have the advantage that every term in it is positive.

127. The Growth of the Partition Function

The partition function p(n) satisfies the inequality

$$p(n) \le e^{c\sqrt{n}}, \tag{1}$$

where $c = \pi\sqrt{2/3} = 2.565\cdots$.

Proof. In our argument we shall use the inequality that if $0 < x < 1$ then

$$e^{-x}/(1 - e^{-x})^2 < 1/x^2, \tag{2}$$

so we prove this first. For $0 < x < 1$ we have

$$e^{-x} < 1 - x + x^2/2 - x^3/6 + x^4/24 < 1 - x + x^2/2 - x^3/8$$

because $1/6 - 1/24 = 1/8$. Consequently,

$$x^2 e^{-x} < x^2(1 - x + x^2/2 - x^3/8) = x^2 - x^3 + x^4/2 - x^5/8$$
$$< (x - x^2/2 + x^3/8)^2 < (1 - e^{-x})^2,$$

so (2) does hold.

Turning to our proof of (1), we shall apply induction on n. Inequality (1) clearly holds for $0 \le n \le 8$, say, with equality only for $n = 0$. (The gap between the values of $p(n)$ we have enumerated and the upper bound in (1) is clearly widening rapidly, so we are free to assume that n is large – not that we need it.) To prove the induction step, recall the recurrence relation

$$np(n) = \sum_{v=1} \sum_{k=1} vp(n - kv) \tag{3}$$

we have proved in the previous problem. In the double sum on the right, the values of k and v are taken to satisfy the condition $kv \le n$. Equivalently, k and v may take any values, since $p(n) \ge 1$ if and only if $n \ge 0$; otherwise $p(n) = 0$.

Assuming that $n > 1$ and (1) holds for smaller values of n, by (3) we find that

$$np(n) \leq \sum_{v=1}^{\infty}\sum_{k=1}^{\infty} ve^{c(n-kv)^{1/2}} < \sum_{v=1}^{\infty}\sum_{k=1}^{\infty} ve^{cn^{1/2}-ckv/2n^{1/2}}$$

$$= e^{cn^{1/2}} \sum_{k=1}^{\infty} \frac{e^{-kc/2n^{1/2}}}{(1-e^{-kc/2n^{1/2}})^2},$$

where in the first summation we take $kv \leq n$. Applying (2) with $x = kc/2n^{1/2}$, and recalling Euler's solution of the Basel problem that $\sum_k 1/k^2 = \pi^2/6$, we find that

$$np(n) < e^{cn^{1/2}} \sum_{k=1}^{\infty} \frac{4n}{c^2 k^2} = ne^{cn^{1/2}},$$

proving (1). $\qquad\square$

Notes. The result in this problem was proved by Erdős in 1942. As he remarked, a similar argument gives a lower bound, so the two bounds together tell us that

$$\log p(n) \sim c\sqrt{n}, \tag{4}$$

where $c = \pi\sqrt{2/3}$.

When Erdős published his paper, this was already an old result: the aim of Erdős was to give an elementary proof of a much weaker version of the great theorem Hardy and Ramanujan proved in 1918. An easy consequence of the Hardy–Ramanujan theorem is the asymptotic formula for $p(n)$ that

$$p(n) \sim \frac{1}{4n\sqrt{3}} e^{c\sqrt{n}}. \tag{5}$$

In his 1942 paper, Erdős went way beyond (4) by giving an elementary proof of the following weaker version of (5):

$$p(n) \sim \frac{a}{n} e^{c\sqrt{n}},$$

where a is a constant, but he could not show that $a = 1/4\sqrt{3}$. An interesting consequence of the elementary argument in this problem is that it explains the mysterious $\pi\sqrt{2/3}$ in the formula for $p(n)$: this constant is a simple consequence of Euler's classical result that $\sum_k 1/k^2 = \pi^2/6$.

By the beginning of the 20th century there was much interest in the growth of the partition function. Calculating by hand, Major MacMahon showed that $p(20) = 627, p(50) = 204,226$ and $p(80) = 15,796,476$, so it was clear that this function growths rapidly, but no pattern could be discerned. The problem of pinning down $p(n)$ with a reasonable accuracy acquired the reputation of being a formidable problem. It was not until 1918 that Hardy and Ramanujan published their sensational asymptotic formula for $p(n)$, whose error term *decreases to 0 as* $n \to \infty$. They proved this by their fundamental method, which was further refined by Hardy and Littlewood and used for several other problems, so that it became known as the Hardy–Littlewood Circle Method.

The Hardy–Ramanujan theorem gave a series for the approximating $p(n)$; this series consisted of pretty complcated terms, but its accuracy was staggering. As Hardy and Ramanujan wrote in their paper:

> *A final question remains. . . . we may reasonably hope, at any rate, to find a formula in which the error is of order less than that of any exponential . . . or even bounded. When, however, we proceeded to test this hypothesis by means of the numerical data most kindly provided for us by Major MacMahon, we found a correspondence between the real and the approximate values of such astonishing accuracy as to lead us to hope for even more. Taking $n = 100$, we found that the first six terms of our formula gave $190\,568\,944.783 + 348.872 - 2.598 + .685 + .318 - .064 = 190\,569\,291.996$, while $p(100) = 190\,569\,292$; so that the error after six terms is only .004.*
>
> *These results suggest very forcibly that it is possible to obtain a formula for $p(n)$, which not only exhibits its order of magnitude and structure, but may be used to calculate its exact value for any value of n. That this is in fact so is shewn by the following theorem.*

After this, Hardy and Ramanujan proceeded to state their great theorem which indeed gave a series which, after only a few terms, determined the exact value of the partition function.

In 1937, Rademacher improved the Hardy–Ramanujan theorem: for every n he found a series *converging* to $p(n)$. Then, in 1943, he gave a substantially simpler proof by constructing a new path of integration to replace the circle carrying first introduced by Hardy and Ramanujan.

References

Erdős, P., On an elementary proof of some asymptotic formulas in the theory of partitions, *Ann. Math. (2)* **43** (1942) 437–450.

Hardy, G.H. and S. Ramanujan, Asymptotic formulae in combinatory analysis, *Proc. London Math. Soc.* **17** (1918) 75–115.

Rademacher, H., On the partition function $p(n)$, *Proc. London Math. Soc.* **43** (1937) 241–254.

Rademacher, H., On the expansion of the partition function in a series, *Ann. Math. (2)* **44** (1943) 416–422.

Uspensky, J.V., Asymptotic formulae for numerical functions which occur in the theory of partitions, *Bull. Acad. Sci. URSS* **14** (1920) 199–218.

128. Dense Orbits

There is a bounded linear operator T on the classical sequence space ℓ^1 such that for some vector $x \in \ell^1$ the orbit $\{T^n x : n = 1, 2, \ldots\}$ is dense in ℓ^1.

Proof. Let $(e_i)_{i=1}^{\infty}$ be the canonical basis in ℓ_1 and $Z = \{z_1, z_2, \ldots\}$ a countable dense set of vectors. Note that if $Z' = \{z_1', z_2', \ldots\}$ is such that $d(z_i, z_i') = \|z_i - z_i'\| \to 0$ as $i \to \infty$ then Z' is also dense in ℓ^1. Let us define two bounded linear operators on ℓ^1. First, T is twice the left shift: $T\left(\sum_{i=1}^{\infty} \lambda_i e_i\right) = 2\sum_{i=2}^{\infty} \lambda_i e_{i-1}$; second, S is half of the right shift: $S\left(\sum_{i=1}^{\infty} \lambda_i e_i\right) = \frac{1}{2}\sum_{i=1}^{\infty} \lambda_i e_{i+1}$.

Let us draw attention to some trivial properties of these operators. First, $\|T\| = 2$ and $\|S\| = 1/2$; in fact, $\|Sx\| = \|x\|/2$ for every $x \in \ell^1$. Second, $T^n e_i = 0$ for $1 \le i \le n$. Third, $T^m S^m$ is the identity for every m, and so $T^m S^n = S^{n-m}$ whenever $m \le n$.

In order to define our vector x whose orbit we shall show to be dense, let us set $n_0 = 0$, and define $n_1 < n_2 < \cdots$ one by one such that

$$\|S^{n_j - n_{j-1}} z_j\| < 2^{-j} \qquad \text{for every } j \ge 1.$$

Since $\|S^{n_j} z_j\| < 2^{-j}$, the sum $\sum_{j=1}^{\infty} S^{n_j} z_j$ is convergent to a vector of norm at most 1; this is our vector x:

$$x = \sum_{j=1}^{\infty} S^{n_j} z_j.$$

We claim that $T^{n_i} x$ can play the role of z_i', so the pair (T, x) will do. Indeed,

$$T^{n_i} x = T^{n_i}\left(\sum_{j=1}^{i-1} S^{n_j} z_j + S^{n_i} z_i + \sum_{j=i+1}^{\infty} S^{n_j} z_j\right)$$

$$= z_i + \sum_{j=i+1}^{\infty} T^{n_i} S^{n_j} z_j = z_i + \sum_{j=i+1}^{\infty} S^{n_j - n_i} z_j.$$

Consequently,

$$\|T^{n_i}x - z_i\| = \|\sum_{j=i+1}^{\infty} S^{n_j - n_i} z_j\| \le \sum_{j=i+1}^{\infty} \|S^{n_j - n_i} z_j\|$$

$$\le \sum_{j=i+1}^{\infty} \|S^{n_j - n_{j-1}} z_j\| < \sum_{j=i+1}^{\infty} 2^{-j} = 2^{-i},$$

so $\left(T^n x\right)_1^{\infty}$ is indeed dense in ℓ^1, as claimed. $\qquad\square$

Notes. This problem was on my Part II Linear Analysis Examples Sheet in 1979. Although it was not one of the most difficult questions, only one student, Charles Read, managed to do it. In fact, he did it easily, handing in his solution at the first opportunity. It is not inconceivable that this was the beginning of Read's fascination with rapidly increasing sequences, a 'trick' he used in his ingenious constructions of operators without non-trivial closed invariant subspaces.

Charles John Read (1958–2015) came up to Trinity College, Cambridge in 1976 as our top scholar. Throughout his undergraduate career, he was most inventive, but somewhat unruly. He was a wonderful, colourful research student of mine, who was always impatient to get his results. As a fully fledged mathematician, he was first a Fellow of Trinity and then a professor at the University of Leeds. Tragically, he died while on a run in Winnipeg soon after he had arrived from England for a research visit. Losing one of my favourite former students and a close friend was a terrible blow to my wife and me.

Read worked much on the Invariant Subspace Problem on Banach Spaces, and solved it completely independently of Per Enflo, whose work, unfortunately, took years to get published. Later Read greatly simplified his own proof and also extended it, first by himself, and then with Gallardo-Gutiérrez in a paper published posthumously. Sadly, soon after his first solution, several unfair comments were published about him, based on false information and malice.

References

Enflo, P., On the invariant subspace problem in Banach spaces, *Acta Math.* **158** (1987) 213–313.

Read, C.J., A solution to the invariant subspace problem, *Bull. Lond. Math. Soc.* **16** (1984) 337–401.

Read, C.J., A short proof concerning the invariant subspace problem, *J. Lond. Math. Soc.* **34** (1986) 335–348.

Gallardo-Gutiérrez, E.A., and C.J. Read. Operators having no non-trivial closed invariant subspaces on ℓ^1: A step further, *Proc. Lond. Math. Soc. (3)* **118** (2019) 649–674.